# Into the Dark

# Into the Dark

*What darkness is and
why it matters*

JACQUELINE YALLOP

ICON

Published in the UK and USA in 2023 by
Icon Books Ltd, Omnibus Business Centre,
39–41 North Road, London N7 9DP
email: info@iconbooks.com
www.iconbooks.com

Hardback ISBN: 978 183773 071 1
ebook ISBN: 978 183773 073 5

Typeset by SJmagic DESIGN SERVICES, India.
Printed and bound in Great Britain

*The mystery of what goes on inside the mind of another person becomes terrifyingly impenetrable in the final stages of dementia; twilight to pitch dark at the vanishing line between life and death.*

*Nicci Gerard*

# CONTENTS

# Prologue

I have a distinct memory of the first time I encountered complete darkness. I might have been seven or eight years old, perhaps a little older; I was on holiday with my parents. We were north somewhere, the north of the Lake District or Scotland – that's a detail I don't remember. In those days, we would overnight in bed and breakfasts, rolling up in a small town where the staff at the local tourist information office would ring round a dog-eared list to find us a place to stay. Mum would sit in the car and I'd go with dad to the counter, listening to the woman chat on the phone to would-be hosts. Depending on the place and the time of year, the process of finding accommodation could be lengthy; we might be on page two or three of the list before rooms became available. On this occasion we must have been later than usual arriving at the tourist information, and the search must have been tricky, because it was late and dusk was falling when we edged along a narrow lane towards a remote farm. I remember the white walls of the house standing out from the deep shadow of the trees as we drove into the yard; I remember being hustled into the warm yellow kitchen. This wasn't the usual B&B my dad would choose, a villa with a sign and a prim garden, pink bedspreads and clean bare furniture; this was a place of muddy wellies and pets and unwashed dishes. We'd been shoehorned into someone's life here, presumably for extra cash. This was a real farm, rundown, busy, and would we like to watch the boys go after the rats, we were asked. And they did go after them as soon as night came, with torches and Jack Russells and sharp spades, the yard a squeal of noise and blood, bare bulbs in the dark.

As part of this ad-hoc deal, I was given a small bedroom at the end of a long landing, another child's bedroom, with teddy bears and books, an old-fashioned narrow bed with a metal frame, and a ceiling painted with stars. There were clothes folded on a chair, others hanging loose from the wardrobe. There was a walking stick propped behind the door, socks around, as there always are. These walls, too, were yellow. I can picture it very well, this room I was stealing for the night from another child, even though I must have only spent a matter of minutes there settling into bed before mum and dad went to wherever they were sleeping. I don't remember waking there the following morning; I don't remember anything of breakfast or what happened next or where we went.

But I do remember the dark.

The light was switched off, the door closed, mum and dad went away, and there was nothing. For the first time, nothing.

I held my hand in front of my face and it wasn't there. Swish, wave, fist, nothing. The socks and the stick, the wardrobe, the window, the yellow, all lost. I could feel the dark, a thing in the room. A thing of substance.

I don't remember being afraid. But the dark was now real to me.

There, in a farmhouse I can't name or place, in another child's bedroom that probably no longer exists, that might have been where it started.

Since then I've taken notice of the dark. I've grown to know a delicate dusk in my fingers like cool cloth, and the full-grip darkness of night, sticky, burring to my skin and my hair and my clothes like goosegrass. I've ventured in moonless drizzle, shapeshifting.

The dark has felt weighted, waiting. But it's still been mostly a kind of aside. It wasn't a thing that mattered.

Then something happened to my dad.

A couple of years ago, an ambulance came to take him to hospital where he was diagnosed with delirium and, subsequently, with dementia. The delirium kept him busy, coming and going from the nurses' station, helping out. He'd worked for the ambulance service for many years and he simply thought he was back on duty again. He was comforted by the sepia familiarity of old habits. But when he came home, his relationship to the dark, as to much else, had changed. There was a metaphorical element to this: the world reduced and blackened, darkness began to hound him, the inevitable closure of night over day. But the difference wasn't only in the figurative sense. He also demonstrated a physical sensitivity to the dark. His illness manifested in an obsession – a bodily preoccupation – with darkness. Early on, he became acutely susceptible to sensory stimulation, to hot and cold, sound, the widening or tightening of space, and particularly to light and dark. He wasn't noticeably afraid of the dark so much as now constantly attuned to it. He became newly alert to shadow. Boundaries between light and dark – the visual transition between sunshine and shade, for example – perplexed and disorientated him, but also transfixed him. He would stand, motionless, and stare at the dark, seeing something, feeling something, the rest of us could not. In unlit corners of the house, he inhabited a kind of new dimension in which darkness allowed for a different configuration of time and space: a sneaky elf-man peered out from behind the sideboard most nights; his parents came and went, briefly spotlit; he heard and smelled a world of things hidden by the dark but immediate and vivid. As the illness got worse, he needed the certainty of constant light levels, an unchanging environment that didn't surprise or confuse him, a monotony of time and place which left no room for the multiplicities of darkness.

Watching dad took me back to the farmhouse bedroom. For him now, as for me then, this was a raw, new, demanding dark, strange and tangible. From the start, inevitably, it had the better of him. This dark was cunning and lithe. As he edged towards it, it circled back on him; it was deeper and darker than he could ever have imagined. There's no way of holding back the dark. It slips through your fingers, pools beneath your feet.

My dad never talked about, or even admitted, his dementia. He didn't ask for help. Perhaps he knew that I couldn't help him, that no one could. Perhaps he didn't want to draw me to that place he was in. But watching over dad's shoulder, I became newly aware of darkness and what it might be; I began to think about it more fully. What had been little more than a careless fascination seemed newly urgent. I had questions: What does it mean to experience the dark? What even *is* the *dark?*

With dad by my side, I set about trying to find out.

CHAPTER 1

# New Moon

Consider utter darkness. Imagine it. It's not as easy as you might think. It's not a matter of just closing your eyes. Even if you screw up your lids, light leaches. There are dots, shadows, colours. The imprint of the lit world blotches and bleeds.

Turn off the lights then. That's better; that's a start. But it's not that simple, is it? At first your unlit room feels properly dark but before long, a shape emerges, a texture. Light squeezes behind the curtains, from the street or the moon, from cars or houses. Your phone glows, or the little red dot on the charger; the digital clock pulses. The room regathers itself, settles, solid and predictable.

This is not darkness. It only hints at it.

Start again. Take a deep breath and imagine somewhere darker. Drop deep underground, a mine, a cave, labyrinthine, unlit. Shuttle out into deep space, a nook between the stars, a universe beyond reach of the sun; drop down deep in the ocean. Can you feel what it might be like, this state of darkness? Does it frighten or console you?

Absolute darkness. Being and not-being. All and nothing. Presence and absence. What kind of thing is it, this real dark?

★⁺₊★ ☾★⁺₊★

Let's begin with Isaac Newton. In the 1660s he turned his attention to the nature of colour, and to the question of what the dark might be and how – or in fact whether – we could see in it. Through a series of experiments using prisms to direct sunlight, he identified the rainbow colours that make up the visible spectrum. He also concluded that darkness is an absence of light. This might seem obvious to us today, but it was a new way of looking at the dark, which until then had largely been considered to be a pre-existing force separate to light and functioning alongside it, a thing in its own right rather than an absence.

Newton's findings answered some important questions, but also posed another. If darkness is a non-existence – the not-being of light – then how do we experience it? Absences, by their nature, don't transmit any energy or any other positive force. No light, no sound, no fluff or dust or feathers. So if there's nothing for our senses to pick up on, how can we say we're encountering it? Most of us would probably say that we know what darkness looks and feels like; we know, in turn, how it makes us feel. But the conundrum remains: how can we claim to witness something that is not there?

This is a puzzle which has exercised philosophers for centuries. The seventeenth-century English thinker and physician John Locke, for example, spent time considering the ways in which we perceive the world, and fumbled around for an explanation of how we can see anything which does not reflect light; that is, how we can see the dark. He decided that 'one may truly be said to see Darkness' because we're able to recognise it as a distinct entity: 'the idea of black is no less positive to one's [mind],' he claimed, 'than that of White.' But there are plenty of people who disagree with Locke. Some philosophers suggest that all we can do is *infer* the dark from our inability to see objects. They claim that we can't actually experience it; all we can do is guess at it. So if we fail, for example, to see a chair in a dark room and fall over it, then we infer that the room is dark and therefore we're experiencing darkness. Others point out that an inability to see doesn't always mean it's dark: when we emerge from a shady tunnel into dazzling sunshine, we can't see, but this doesn't mean it's dark. In fact, quite the opposite.

Generally, both philosophers and physicists like to suggest that reality is positive, and that negative states can ultimately be explained in terms of positive ones. So, for example, defining the dark side of the moon as the part of the moon's surface that doesn't reflect sunlight, is really about where the light is, rather than where it isn't – it's about the positive effect of light rather than the negative state of its absence. But, of course, the moon continues to be unilluminated in parts, no matter how we describe the bits we can't see. The twentieth-century French playwright, Jean-Paul Sartre, suggested that absence depended on a positive expectation in our minds: we can only note that Pierre is absent from the café if we expected to see him there in the first place. And again, the explanation is finally positive – just because Pierre is not at the café, doesn't mean he's been sucked off the face of the earth. He's probably at home in bed.

Under these terms, darkness only exists as our minds process the expectation of light. It's a deficiency, a disappointment. It's something missing, that's all. But is this your experience of the dark? Moonless at midnight, if you walk the fields or the streets, does the dark around you fall flat and lifeless? Is it just no-light? Or does it seem to have body and essence; does it seem material, tangible even?

Centuries of mind games, investigation and practical experiments have done little more than prove that darkness is an anomaly, a complicated phenomenon which is intrinsically connected to our sensory perception; a non-existent state which most of us would claim to experience as a real thing. But there is one matter on which most science and philosophy agrees: that our concept of the dark is linked to optics. We mostly experience darkness through sight. At a basic level, we know it's dark because we see that it is. Although even this apparently simple assumption is fraught with complications and doubt. Is it even possible to see the dark? This is not a question of whether we can see *in* the dark but of whether we can see the dark itself, whether it's possible to see an absence, something that isn't there. Which is different again from not being able to see at all. A blind person can't see the darkness of a cave, for example – rather they fail to see anything. But this does not necessarily drop the blind person into the dark: when the BBC reporter Damon Rose lost his sight, he said that what he most missed was darkness; he was plagued instead by the constant distraction of changing brightly coloured dots and splodges that he termed 'visual tinnitus'. To *see* the dark – and know that it's dark – we have to recognise it as a state that exists outside us, a thing which our eyes can witness, sending messages to the brain to confirm that, therefore, darkness exists.

Although even such an explanation may be a bit old-fashioned: contemporary theories of consciousness would propose that I've got things the wrong way round, so that while we've been habituated over centuries to thinking of signals pouring *inwards* to the brain from our senses, in fact, the action is reversed so that the brain scans *outwards* in an attempt to keep us safe in our environment. In this scenario, the brain generates a series of informed conjectures about the world around us by processing the feedback it gleans to give form and thing-ness to our environment. 'Our perceptual world alive with colours, shapes and sounds is nothing more and nothing less than our brain's best guess of the hidden causes of its colour-less, shapeless, and soundless sensory inputs,' claims the neuro-scientist Anil Seth, 'Normal perception – in the here and now – is indeed a case of controlled hallucination.' In which case, the dark, like my dad's dark, would not come in from outside, through the eyes to the brain, but would be a construct of our brain projected outwards to help us make sense of what's around us. Which raises some troubling questions: what does it mean if dad's brain, a dying brain, increasingly fashions its world as a thing of darkness? Why might our conscious self place us in the dark? My inquiry into the dark will bring us back to the brain again and again; we'll see how closely darkness and consciousness intertwine, how they dodge and scuffle. Is darkness outside us or within, physical or psychological, real or imagined? We'll come back to all these questions. But for now we need to pin down the discussion of optics.

At this point, it might be useful to offer a biological explanation: darkness is the absence of photons in the visible spectrum as perceived by the retina. That is, where there's no visible light for our eyes to process, there is the dark. Most of us will respond to light wavelengths of between 380 and 750 nanometres, from violet light at the shortest end of the spectrum to red at the longest. All electromagnetic radiation is light, but we can only see a small proportion — cone-shaped cells in our eyes act as receivers tuned to the particular wavelengths in this narrow band. Other areas of the spectrum have wavelengths too large or too small and energetic for the biological limitations of our perception. So, while there may be light bouncing around beyond the visible spectrum, as far as we're concerned, if it's not in the rainbow of perceivable colour, then our eyes tell us that we're in the dark.

This makes sense. We know that if we turn the lights off in a mine shaft hundreds of feet underground, we're going to be thrown into darkness. Even if we've never done such a thing, we can imagine what the experience might be like. It feels instinctive knowledge. And we have historical and anecdotal evidence for it.

The first mining manual, *De Re Metallica* by Georgius Agricola, originally printed in Latin in 1556, drew the reader's attention to 'the dead darkness' of the mine which, Agricola suggested, was inherently disturbing, a place naturally inhabited by ghouls and phantoms, lending itself to 'the substantiation of the supernatural'. Accounts from children working in mines throughout Britain during the nineteenth century consistently return to mention of the dark. They never got used to it. It intimidated and frightened them through the long dirty shifts underground. Those whose job it was to man the system of doors and traps were especially likely to spend many hours in a state of black-out which effectively silenced as well as blinded them: 'I have to trap without a light,' reported eight-year-old Sarah Gooder, who worked in the Gawber pit near Barnsley in 1842, 'I'm scared ... Sometimes I sing when I've light but not in the dark; I dare not sing then.'

In 1855, a group of curious ladies and gentlemen took an excursion a thousand feet underground into a deep pit in the Eastern Virginian coal field near Richmond, USA. They thought they'd known what to expect, what they were letting themselves in for. They believed they could picture the dark. But when they dropped 300 metres underneath the rock, the darkness overwhelmed them. Their senses failed. In this 'darkest abode of man', one of the ladies had a fainting fit and had to be helped to the surface; a male adventurer noted unnatural sweats and dizziness. Several of the party held hands to anchor themselves because they felt as though they were floating away. Time and space were upended and, perhaps most dangerously of all, the sense of self that sustained them above ground was slipping. Distinctions were meaningless in the dark; the wealthy day-trippers were losing themselves among the slaves who worked the mines. It turns out they hadn't imagined the darkness at all.

As these reactions suggest, darkness is more than a case of not-seeing. So what is it? A feeling, an experience, a change of state? Biological explanation of the dark is not even the beginning. 'What is the meaning of having one's body *full* of darkness? It cannot mean merely being blind,' asked the nineteenth-century polymath, John Ruskin, who spent his life preoccupied with ways of seeing. It's an interesting phrase he chooses: one's body full of darkness. *Full.* Replete or overwhelmed. Dark as a thing that can inhabit or pervade us, belong to us. What, he wonders, is the meaning of such a thing?

Art, literature, physics, medicine, religion, psychology, philosophy – they all approach and examine the dark and yet there are all kinds of things that we still don't grasp. Like the deep oceans, the dark sits below and around us, unknowable. We remain captivated, baffled and appalled. I came to this investigation because I love the dark, the shifts and folds of it, the feel of it, like the unspooling of imagination, a non-feeling, an intimation, the brush-breath of an insect wing on skin. I love the pierce of stars, the unseen hush of breeze, the disintegration and suspension, the wholeness. But this is only my dark, and only for now. I also took on the dark because of dad's dementia: for him darkness is a menace and a warning. He didn't journey towards the dark; it came for him. Predatory, it stalked him, circled, hunted. If he could, he'd have run from it. He certainly tried. For him, darkness is a terrible blackening of the world, a gobbling-up, a disappearing. This was a darkness I couldn't experience, and could not come close to. It was dad's only, a terrible thing to own. But in looking into the dark, I could hold his hand for a while.

Darkness is shifty, a thing of many textures and many moods, a state of fascination and of horror, an absence and a presence, solace and threat, behind us and inside us, a beginning and an end. This book is the story of the many darks that fascinate and assail us. It faces the darkness full on in all its guises and mysteries, celebrating it as a thing of beauty, peering into the void.

★⁺₊★ ☾★⁺₊★

Consider again, then, the question of utter darkness.

The twentieth-century American philosopher, David Lewis, took on the riddle of whether we could see in the dark, and whether this meant we could see the dark itself. His conclusions were ambivalent. On the one hand, he pointed out that it's only our eyes that can tell us whether it's dark or not. We can't tell by smell or taste; we rely on sight to confirm darkness, 'in the pitch dark, we find out by sight that it is dark.' But he also noted how our eyes can deceive us: just because we can't see anything it doesn't necessarily mean it's dark because 'we also do not see in dazzling light or thick fog'. So where does that leave us? Lewis sat on the fence, saying, 'In a sense, we do see in the dark when we see that it is dark. In a more common sense, we never see in the dark.' So darkness remains an enigma, there and not there, seen and unseen.

Why is it so difficult to pin down? One reason is that our brains rage at the dark, evade it, barricade us against it. If you were shut deep in a mine, your brain would reject the darkness and fool the eyes into 'seeing'. It would rummage around among your memories to create forms, shades, patterns. It might take a little while, but in time it would reconstruct a world without the terror of utter darkness; with a nifty VR simulation, it would allow light.

In 2009, the artist Alan Smith took a party of six non-cavers into the disused lead mines underneath Nenthead in the North

Pennines. These largely intact old workings date back to the eighteenth century and stretch for around 1,500 square kilometres under the hills. Narrow tunnels and shafts lead from one mine to the next; at its peak, a network of 200 mines channelled people and materials through the rock, creating a huge dark world like a negative imprint of the one above ground. Smith's party spent twenty-four hours exploring the underground system, bedding down overnight in the cavernous 'Ballroom Flat', a vast hollowed-out void named after a bizarre Masonic dinner party which was held there in 1901.

I've been into the mines at Nenthead; I've splashed, crawled and tripped through the levels to the ballroom cave. It's a strange, exhausting experience. Worming towards the belly of the earth alerts both the mind and body; both buzz with excitement; both strain to hold the experience in check. When you reach the ballroom, turn out the lamps on the helmets and douse the torches, the darkness rampages, magical and monstrous. It hits you with force like a wave breaking over you in the sea. It's thirty years since I disappeared into the dark of the Ballroom, but I remember it clearly: such darkness is unforgettable. I remember the feeling of my body falling from me with the sudden stubbing out of light, a feeling of weight and weightlessness. My mind shrank from the dark, cowered; leapt into it, opening out into this new dimension as though my consciousness had suddenly sprung huge, limitless. It was only a matter of moments before the torches were switched back on, but it felt as though in those moments I had been both annulled and realised.

The dark nurtures paradox. Certainties of thought and experience hold little sway as our minds renegotiate the substance of reality. During the night of Alan Smith's expedition into the mines from Nenthead, one of the participants first noted her heightened sensory awareness, 'the distinctive and varying textures of things in the dark'. Then came something stranger, the warping of time – 'One hour could become one minute, one second, one nanosecond. I am floating in time without measurements'. And finally, as her brain worked to negate the darkness, she started to 'see' things:

> I … wake to total blackness. No candles. Opening and closing my eyes makes no difference at all. Am I even awake?... I see short yellow lines emanating from a black void. It seems to turn into an opening that I might rise up and go through … slowly drifting blobs of varying size, shape and transparency; tiny bright dots moving rapidly along squiggly lines; subtle bowtie and hourglass shaped patterns. You can see the blood vessels in your own eye appear like a tree.

This experience of sight-in-darkness, often described as 'retinal noise' or 'light chaos', is not unusual, and is very similar to what Damon Rose described in his account of blindness. It's a strange kind of etch-a-sketch sensation where the dark is scratched and blotted with shifting clouds of floating spots, or swirling ribbons of light.

These forms and patterns can seem to have substance and grain so that if you try to move around in a dark room, the retinal swirls can become confused with real objects. In a blackened space or during a night-time walk, our eyes and our brains collude to refute darkness, to disempower it, and to replace it with more familiar and identifiable 'sights'. So the colours and shapes imprinted on the dark can take on recognisable form, tricking us into mitigating the discomfort of the void through our own invention. In 2007, the German artist Marietta Schwarz wore a blindfold for 22 days, and gave an account of everything she 'saw'. Her brain activity was tracked by an MRI scanner, and studied by researchers at the Max Planck Institute for Brain Research in Frankfurt. During her three weeks of enforced darkness, she described seeing leopard-print patterns and the opening credits of *Star Trek*. The scan showed her brain was lighting up to display the exact same activity as it would have done under 'normal' vision conditions. That is, it was seeing these things. It was not imagining the sight of the Starship Enterprise; it was actually seeing it. When we move in the dark – waving a hand in front of our face, for example – this creates sensory signals which create real visual perceptions in the brain, even in the complete absence of optical input. Our brains simply will not accept that they can't see anything: they challenge the fact of darkness; they blot and shift it, doing their utmost to reconfigure it. By doing so, they reclaim it for experience, reinventing reality as shifting, negotiable, independent of sight and light. Darkness forges a new sense of what is real.

In 'We Grow Accustomed to the Dark', the poet Emily Dickinson describes this tussle with the dark, investigating how we 'grope' and 'adjust' as an extended metaphor for change and resilience, for the process of overcoming adversity. She captures our brain's natural elasticity in the image of reconditioning darkness:

> The Bravest – grope a little –
> And sometimes hit a Tree
> Directly in the Forehead –
> But as they learn to see –
>
> Either the Darkness alters –
> Or something in the sight
> Adjusts itself to Midnight –

We adjust ourselves to midnight: this is a hopeful poem. It puts a positive spin on the shifting of the dark. Our brains negotiate the dark to overcome it; they manipulate image and memory with such mastery that we don't even know the lights are off.

Why be afraid of something our brains can bypass and re-work? If we can adjust to midnight, why worry about it?

Theories of quantum mechanics might also allay fears of the dark. Quantum mechanics says no space is ever dark, not even the darkest cellar or deepest mine. The physics draws attention to the omnipresent fluctuations of light that exist beyond the visual spectrum, vacuum fluctuations or light noise which interact with atoms to produce photons – bundles of electromagnetic energy. Since these photons exist everywhere, they disrupt darkness everywhere. For quantum physicists, true darkness – the complete absence of photons – is impossible. There will always be the un-dark of photon activity. So even if you were stuffed in a cloth hood and placed in a black box in outer space, quantum mechanics would posit that you were not in the dark at all because photons would still be detectable.

Recent findings from the NASA New Horizons space mission suggest that you don't even have to rely on the tiny photons of quantum physics to keep the dark away. The mission set off to study Pluto and its moons in 2015, then just kept sailing on – it's now more than 4 billion miles from Earth, somewhere way out in a place you would think of as dark. The photos sent back from the mission were taken far beyond any sources of light contamination, such as the dust particles in the inner solar system that are lit up by the sun. But when you study these photos and remove all known sources of visible light – from the stars, from the Milky Way, from the camera itself – and then any light that could be attributed to known or suspected galaxies, then, after all that, they're still not dark. The amount of light leaking from unidentified sources is not a glimmer or a trickle: it's a definite glow, more or less equivalent to all the light coming in from known galaxies. It may be an indication of other, unrecognised galaxies waiting to be discovered, or of another unknown source of light – no one knows yet. But whatever the reason, it leaves us still in the not-quite dark. The New Horizons evidence supports quantum physics in suggesting that complete darkness doesn't exist.

But the contention that darkness is impossible doesn't tally with our experience, does it? Or it misses the point. Darkness is difficult to fully imagine but we've all seen it, felt it, run from it. Legend and literature assert that darkness is real; film has it creeping up behind us. It's integral to our understanding of day, season, life, death.

And if darkness doesn't exist, what is it we're trying to flee?

★⁺₊★ ☾★⁺₊★

I'm at home, walking near the house at night. I'm in the woods in Wales, in the narrow scrape of valley, the cwm. My hands and feet are tingling cold. I can hear the stream rushing, singing over the stony ridges, a rustle in the wet leaves on the ground. Damp air prickles like the dark on my skin.

The January new moon rose this morning just after five and set hours ago, in the mid-afternoon; unfastened from the dark, it was a day-moon, tracking alongside the sun. It was invisible. In a few days' time, as the lunar phase develops, I'll easily be able to see the thin crescent slice of a young moon in the evening, but for now there's nothing. Each cycle of the moon begins in darkness.

I find my way in these woods by memory or instinct, by the vegetal smell, by touch, stumbling, but not by sight. I'm a long way from streetlights, a road, a house. I have no torch. If I hold out my hand, I can't see it. If I look down to my feet, they are not visible. It's possible I'm not here. Quantum physics tells me that I can't really be in the complete dark, but this encounter refutes the niceties of science.

This is a small steep slice of woodland, mostly ash, sycamore and holly, a few huge old oak trees. The stream runs quick and noisy. On the far side, the bank rises precipitously into gorse; on this side there's an area of stony level ground surrounded by some of the tallest trees. I am far out west where the slatey hills crumble and slide towards the sea. This is a place I know well. I've been here many times in daylight; I've sat by the stream, watched the squirrels and woodpeckers in the trees, the buzzards and red kites above. It's a sheltered, safe valley that cuts directly down to the coast, private land. It's most unlikely that there are ne'er-do-wells hiding in the shadows. Dragons, these days, are rare; the only animals I should encounter are foxes, badgers, toads, cats. There's little danger of landslide, sinkhole, hurricane, or lightning strike. But none of the things I know about these woods hold true now. I can't see any recognisable landmarks; I can't make out the contours of the land; the trees could no longer be there, for all I can tell. This is not the place I know by day. Darkness has transfigured it. The familiar has, literally, disappeared leaving a half-known place, a place that has warped and buckled, distances longer (shorter?) than I'd recalled, trees out of line, dips and rises where they shouldn't be, threat where I'd never conceived of it. As it is, I relish the blindfold challenge. Dread, fear, disorientation are momentary, indulged, invited. A buzz. I enjoy the ethereal feeling of being in the utter dark, of floating – or is it sinking? – untethered. But still I think – what if I'd been lost in the darkness? Really lost. Here, in the pitch black. What if I'd been where dad is?

Pitch. A word that's been used for over seven centuries, since around 1300, to help us reach for the dark. The phrase 'pitch black' was coined in a 1598 satire, *The Scourge of Villanie*; the diarist Samuel Pepys describes coming home after midnight in January 1666, when everything was 'dark as pitch'; Daniel Defoe used 'pitch-dark' in 1704 to describe conditions during a hurricane which had devastated parts of London the previous year. Pitch is the gummy, resinous residue from the distillation of wood tar or turpentine, commonly used as waterproofing for sailing ships. We don't waterproof many sailing ships these days so our experience of pitch largely exists in metaphor. But it's clung on, sticky, because we can't find a better way of explaining what it is that threatens us. 'Pitch dark' is about feel as much as sight. It reminds us that dark is viscous, tangible, impenetrable. We get stuck to it like flies on gummy paper.

Dark. A word without precedent, without patina. No one really knows where it comes from. It's unique, strange. It edges towards us with menace. The Old English *deorc* is probably related to the Old High German *tarchanjan,* meaning to conceal or hide, a mystery, a danger. But the roots of *tarchanjan* are unclear; unusually, it doesn't emerge from Latin or Greek. Dark sprung at us from nowhere. And Anglo-Saxon poets sometimes chose another word: *Þeostrum*. An absence of light, conjuring not only a biological darkness, but also the nothingness of being separated from God. Darkness theological as well as physical. The loneliness of being in the dark. The terror of absence. Dark was no longer only a description of night but a mood, a behaviour, an alienation, a belief. It had become a loaded word, a concept, a puzzle.

The word for dark struggles to contain the multiplicity and slipperiness of the idea. The Greek *σκότος* (skotos) is both literal and metaphorical, indicating both the black of night and the gloom of misery. European Romance languages – French, Spanish, Portuguese, Italian – use a variation of the word 'obscurity' to include notions of concealment or estrangement. The Chinese noun 黑暗 leads us into moral murk. Our language of the dark is figurative but still falls short. We struggle to conceive of absolute darkness, and we labour to express it. Taking part in a 'dark residence' at London's Southbank Centre in 2018, the poet George Szirtes attempted to capture the abstraction of the dark, its changeableness, unpredictability, inexpressiveness:

> The idea of total darkness is not the same as total darkness.
> The idea of light is not the same as light.
> The words expressing the idea of light or total darkness are not ideas.
> This word may be imagined vanishing into total darkness.
> This word has begun to express an idea but most of it is lost in darkness
> This sentence is not total darkness.
> This one is

Poets have grappled with similar ideas for centuries. Take Byron's marvellously menacing post-apocalyptic poem 'Darkness' (1816). Something has happened. Byron doesn't tell us what, but this is an unbounded, terrible event. The entire world has been overwhelmed and the consequences are catastrophic: 'the bright sun was extinguished' and the earth is 'icy … blind and blackening'. There are no winds, no tides, no moon. What Byron relies on to approach the full idea of the darkness of the title – to try to convey the overwhelming sense of loss, the abyss which has consumed human warmth and hope – is not florid description or complex verse but a series of bald negatives: the stars, he says, are 'rayless' and 'pathless'; the air is 'moonless'. War, starvation and cannibalism ravage the world, where human bones lie about uncared-for and 'tombless'. Everything is reduced to a 'lump of death'. The poem reaches a climax in a glut of absence – 'seasonless, herbless, treeless, manless, lifeless'. Darkness is indescribable, uncapturable, except as a state where things are not. It is a turning upside down of human values and achievements: passions forgotten, hearts chilled, cities annihilated, the natural world decimated. It consumes all the forces of love and loyalty and kindness to become a being, a female figure, all-powerful and overwhelming. 'She was,' Byron says, 'the Universe', but she is shapeless, voiceless, indefinable. Even as a personification, darkness has no form, no body. It remains a lack. It eludes.

A century or so later, in Dylan Thomas' radio play, *Under Milk Wood* (1954), darkness remains elusive (and allusive), a swirling rhythmic darkness of dream and memory, of creeping consciousness, a place of the mundane and the bizarre. Here we are, between the 'silences and echoes of the passages of the eternal night'. Here we are, on the edge of death. Like Byron, Thomas' dark is a state of negatives, 'moonless' and 'starless'. But it's something more. A darkness of small things, felt as much as seen, instinctive and incantatory. 'It is spring,' the poet-narrator tells us, 'moonless night in the small town, starless and bible-black, the cobblestreets silent and the hunched, courters'-and-rabbits' wood limping invisible down to the sloeblack, slow, black, crowblack, fishingboatbobbing sea.'

The sloeblack, slow, black, crowblack, fishingboatbobbing sea. This is one of my favourite evocations of the dark. The town of

New Quay, inspiration for *Under Milk Wood*'s fictional setting of Llareggub, is just a few miles south from my night-time walk. As I leave the woods and the cwm, the path climbs, widens into a stony track, the trees thinning and then giving way to high hawthorn hedges. I can hear the sea, waves noisy on the shore a mile or so away, high tide or thereabouts. The lights of Aberystwyth, a small town, hover now in the distance. I walk towards them and from near the coast path, later, I'll also be able to see the lights of New Quay, clustered in a nook of Cardigan Bay, white and yellow against the sloeblack hills and the crowblack sea.

This is usually how we treat the dark – as a brief annoyance to leave behind or throw off, a moment of discomfort or absence before light becomes visible again to entertain and reassure us. Darkness is an interlude in the usual state of things, a hindrance to be overcome; we battle it. The expressions we use reinforce this sense of skirmish. The dark is frequently 'put to flight', 'overcome' or 'driven out'. It is dispelled. But whatever we do, of course, and whatever words we employ, the dark remains. It lingers, it returns.

It is a moonless winter night on the west coast of Wales and I turn away from the pinpoints of light by the sea and pause for a moment between the hedges. The sea sighs and a stand of pines further along the path hums in the wind but I can't see them. The darkness eases back.

★⁺₊★ ☾★⁺₊★

In 1617, an English physician called Robert Fludd took a page in his book *Utriusque Cosmi* and filled in a large black square from edge to edge. In the margin, he wrote '*Et sic in infinitum*' – the same going on for ever. This was his best stab at representing the beginnings of creation, the darkness before the first moment of light, the nothing before something. Blankness. Un-being. Darkness was there before us and will be there after us. *Et sic in infinitum.*

In 1914–5, at the outset of a terrible war, the Russian avant-garde artist, Kazimir Malevich, painted *Black Square on a White Ground.* Unlike Fludd, he didn't write anything on the canvas. He left his dark context-less, possibly meaningless. It's up to us what we bring to the black; perhaps we're drawn to the simplicity of it; perhaps we're influenced by the connotations we bring: we talk of a miserable 'black mood', of the evil of 'black magic', of the financial disaster of 'Black Monday'. Malevich's black painting became a modernist icon but many other artists – Mondrian, Klee, Serra – were drawn by the dark, by the abstraction of black, by a state without time and place. The absence or concealment of light and colour – often of form, too – in these black works reached towards the abstract nature of darkness, returning us to our inability to adequately imagine, describe or prescribe it.

At the Musée Soulages in Rodez in south-west France, you slip through galleries of black canvasses made by the artist Pierre Soulages from 1946–70. He calls his practice *outrenoir* – beyond black. He gouges, folds and sculpts the dark, which is charged, relentless. For a while, as you move from room to room, you resist such monotone repetition. You think you want something more. But then you pause, you look again, thrown in on yourself, back on yourself, and the dark becomes alive, intimate, emotive, inviting. 'My pictures are poetic objects capable of receiving what each person is ready to invest,' Soulages said.

In the 1950s and '60s, as Soulages was working towards *outre-noir,* the American artist Ad Reinhardt painted a series of black squares which he stipulated should be viewed in low lighting – in the dark. He eradicated any evidence of brushstrokes and used a matt finish to minimise reflections from the surface. This was, Reinhardt claimed, 'the death of painting'. There was nowhere to go from here. Darkness of this nature marked the end of visual representation; there was (literally) nothing to see. Discussing his work, Reinhardt echoed Byron by referring to the black paintings as a collection of negatives: 'a pure, abstract, non-objective, time-less, spaceless, changeless, relationless, disinterested painting'. As with Soulages, this is difficult work which gives nothing away, which refers to nothing, which admits nothing. The brain works to challenge the uniformity and absoluteness of the black surface. What you see changes: blacks slip and slide, adopt hints of colour, blues or reds, which then fade; texture ruffles to the surface. One minute you're looking at a grid or a cruciform, perhaps; the next at a cloudy sky or the hint of a landscape. It's like the tricks played by the eye in the deep dark of the Ballroom Flat in the Nenthead mines – your vision changes, perception changes, you change. The paintings do not.

Our perception of black depends on the circumstances in which we view it. Blacks look blacker when contrasted with white; light is needed for us to perceive the deepest blacks: hence Malevich's use of a white background for his black square. On the other hand, what's known as the 'assimilation effect' can also make a surface look darker when it's seen against a dark background. And the proportions of the background can also play a part: a black shape can be made to look blacker just by increasing the size of the background, since our visual response interprets wider expanses as lighter and so sees the black as blacker in contrast. Depth perception, too, effects how we see black: our eyes can be tricked into seeing blacker than black patches depending on whether they seem nearer or further away. Even invisible contours can create a blacker-than-black effect, which is what Soulages exploits in his work.

Black objects are the best absorbers of energy (and also the best emitters). They are black *because* they absorb energy well; they don't absorb energy because they're black. A black moth on a pale tree trunk appears black because of its energy absorption. A green beetle is green because it absorbs all the wavelengths of light except those associated with green; a black beetle is black because it absorbs all wavelengths of light, without any exceptions.

But black things are not intrinsically dark: black objects need light to become black. A black thing – black paint, for example – needs light to manifest its ability to absorb light. This is different to the dark: darkness, as Newton taught us, is an absence of light; black takes light and changes it.

There are other anomalies. An unlit room in a house can be dark without being black – when the lights are switched on, we might discover bright red walls and yellow furniture. And beyond the polar circles, there can be months of darkness, but this need not always be black: Norwegians sometimes refer to the winter dark as the 'blue time'; more specifically, winter afternoons around 1 or 2 o'clock are referred to as the 'blue hour', when the arctic light reflecting from the sea and snow turns the landscape an evocative deep blue, lending an effect of glassy stillness. This is completely different to the blackness of the deep mine or even a moonless British night. Here, the dark dances and deceives and is a thing of colour.

But mostly darkness is intangible, colourless, lightless. In the end black is the closest we have to representing it in any meaningful way.

In 2014, a chemical company in Surrey unveiled a new product which they called Vantablack. It was, they claimed, 'the blackest material in the universe, after a black hole'. This is the manufacturing of absolute black, an approach to darkness. Absorbing up to 99.965 per cent of visible light, Vantablack is made by chemically growing a forest of skinny tubes, each one 300 times as tall as it is wide. When light strikes, instead of bouncing off, it becomes trapped among the tubes. Unable to escape, it's absorbed, eventually converting into heat. Light is effectively gobbled up. Used scientifically for telescopes and to improve the performance of infra-red cameras, as well as to 'mask' stealth aircraft, the potential of Vantablack to represent – and exude – darkness was also recognised by artists: its potential for creating new ways of using and perceiving black was described by one as 'like dynamite in the art world'. In February 2016, the sculptor Anish Kapoor won exclusive rights to the artistic use of Vantablack in a controversial deal that allowed him to explore notions of the void and negative space but excluded and infuriated other practitioners. Discussing the controversy, Kapoor claimed that it was the very nature of black that set off such strong reactions in his rivals. Black, he said, was 'so emotive... perhaps the darkest black is the black we carry within ourselves ... There's a psycho side to blackness that we don't associate with other colours.'

A psycho side. Darkness within as well as without. Now we're getting somewhere. Four hundred years of black squares suggests how long we've known that dark is not just the night outside the window. If it was, we could close the curtains. Dark is visceral as well as optical, a psychical as well as a physical phenomenon. Paintings which manipulate the apparent nothingness of black are challenging our thoughts and perceptions – our sense of ourselves – as much as, or instead of, our visual sense. Since art has been made and stories have been told, darkness has been entwined with self, with identity and being. The stencilled hands that recur in ancient cave paintings around the world reach out into the dark and grapple with it, the most recognisable of human forms silhouetted against shadow. In Greek mythology, Erebus, the personification of darkness and brother of Night, is elemental, not displaying human shape so much as a swirling, primordial state of being. One of the first five beings in existence, born of the chaos that preceded life, he is integral to the creation story, to an understanding of human place in the world.

If we're to grasp anything of the dark, then, we need to tackle its 'psycho side'. We need to reach for the dark inside us.

Let's begin with the story of Cupid and Psyche. First recorded in the second century AD and retold ever since, the Greek tale tells of the beautiful princess Psyche, who, although much admired, has never found love. When her puzzled father consults the oracle, he's informed that, despite her many advantages, the only husband she can expect is a winged serpent who will terrorise the world with fire. Unsurprisingly, perhaps, such a prediction causes some consternation. Horrified by the prophecy, Psyche's family casts her out, abandoning her on a mountain to wait for her dragon-mate to claim her. This is a moment of darkness, of death – Psyche wears her funeral robes. Transported by Zephyrus, the wind, to a beautiful meadow, Psyche then falls into another dark, into sleep, waking in a magnificent house. Here she stays, alone, looking towards the night. That's when her lover comes to her, in the dark, unseen. She must never try to catch sight of him in the light, he decrees. Passion, belonging, love come with the dark and out of it. But Psyche cannot live with the not-knowing. Encouraged by her jealous sisters, she becomes desperate to allay fears that she's bedding a monster, so she lights a candle one night to catch a glimpse of her strange husband. When a drop of wax falls on Cupid and wakes him, he starts up in anger and flies away.

Psyche, the total human mind, the conscious and the subconscious, from the Greek ψυχή, life, long associated with the concept of the soul and the spirit. But in the story of Cupid and Psyche, Psyche's contentment, her physical satisfaction and her spiritual fulfilment, are dependent on the dark. When she tries to dispel it, she is punished and abandoned. There are many readings of the story from many angles but whatever the emphasis, the dark wriggles through. Our minds, the conscious and the subconscious, are intrinsically tangled in the dark.

Perhaps this is why we so often fall back on language of the dark in our attempts to describe some of our most intense and impenetrable psychological states. We commonly talk about 'dark moods' and 'dark thoughts'; patients trying to communicate the lived experience of mental health conditions frequently draw on an instinctive understanding of darkness to help them explain their feelings. In his memoir *Darkness Visible*, William Styron argued, for example, that 'depression' was simply too bland and indifferent a word for the 'blacker forms of the disorder': the sinister, physical grip of the condition, he noted, was much more akin to our experience of being stranded in the dark. Describing personal experience of mania, the American clinical psychologist Kay Redfield Jamison described the lurch from high to low mood as a transition from the light of 'shooting stars' to 'the blackest caves of the mind' where she felt 'enmeshed' in a series of uncontrollable and difficult emotions. A Canadian doctor described his journey to wellness as 'walking out of the darkness'. It's perhaps not surprising, then, that many of us report a direct link between darkness and our mental health: the association between long periods of darkness and depression is well established, and contributes to our understanding of Seasonal Affective Disorder, a 'winter depression' that feeds on long nights and low light levels. Rats kept in the dark for six weeks have been shown to suffer damage to the brain regions which control mood and emotion.

Since Roman times, the dark has been associated not only with our bleaker moods, but with the shifty unpredictable subconscious, with duplicity and treachery. The Roman scholar, Quintilian, defined 'bad intention' as 'something done … at night'. Medieval towns generally prohibited any transaction after dark because it was deemed naturally untrustworthy; legal contracts and covenants were often considered invalid if executed during the hours of darkness. In recent years, psychologists have identified what they call the Dark Triad, a term coined in 2002 to define all the bad tendencies that lurk within us, the worst elements of the human psyche, singled out, distilled: Narcissism, defined as egotism and a lack of empathy; Machiavellianism, manipulation, self-interest and callousness; and Psychopathy, antisocial, selfish and impulsive behaviour. The triad research aims to order and manage the most difficult elements of behaviour, unearthing those parts of ourselves we try to conceal. Police and the courts sometimes test people for the cluster of Dark Triad personality traits to identify a tendency to criminality; big business will use it to weed out those prone to bullying – or to identify daring risk-takers. But research has also discovered how deep-seated are the darkest parts of our natures. The three traits have been shown to be closely related – most of those who exhibit elements of one, also display elements of the others – and all three are unshiftable, linked to each other and embedded in our humanity by genetics and social evolution. That is, dark behaviours have been proven to be genetically part of our make-up; the development of social constructs and habits over thousands of years has entrenched darkness inside us. There's no escaping it.

The publication of Joseph Conrad's novella, *Heart of Darkness*, in 1902, one hundred years before the 'discovery' of the Dark Triad, told us much the same thing. Like the story of Cupid and Psyche, the narrative can be read in many ways and is the focus of much discussion about its portrayal of Africa, colonialism and race. But as the title suggests, darkness is inescapable. Conrad's narrator is a middle-aged ship's captain called Charles Marlow, a white European who takes his reader-listener on a voyage down the River Congo into an Africa that was, at the time, being terrorised and desecrated under the rule of King Leopold II of Belgium. *Heart of Darkness* helped expose the atrocities taking place in the colony and contributed to an international protest which eventually brought Leopold's rule to an end in 1908. But it's a book about psychological darkness as much as political savagery. It's a story that never sees the light, beginning in a gloomy England – 'one of the dark places of the earth' – and travelling to another 'place of darkness', that is Africa. It is a tale of greed, evil and madness, menace and foreboding and estrangement, of a nightmare state where place, self and time dissolve as Marlow recalls his journey into the continent's interior:

> Going up that river was like traveling back to the earliest beginnings of the world, when vegetation rioted on the earth and the big trees were kings ... We could not understand because we were too far and could not remember, because we were traveling in the night of the first ages, of those ages that are gone, leaving hardly a sign – and no memories.

Darkness here is so complete, it precludes even the half-light of memory, and as the novel progresses, drenched in the unease of shadow, literal and figurative, it exposes the darkness of the psyche that lodges in sorrow and fear, in the selfish, the callous and the impulsive. 'We penetrated deeper and deeper into the heart of darkness,' Marlowe explains, describing a voyage to the underworld in the tradition of Homer's *Odyssey* or Dante, and in so doing, taking us on a journey into the deep recesses of the human condition. 'We live in the flicker,' he says, a brief insubstantial moment of light, although we may, he concludes, be 'too much of a fool … too dull even to know you are being assaulted by the powers of darkness.' *Heart of Darkness* is, said Conrad, 'something quite on another plane than an anecdote of a man who went mad in the Centre of Africa'. It's a journey of self-discovery, to the roots of self, to the losing of self – and hence into the dark.

It's January again, new moon. It's softly, impenetrably dark. The air is still, nutty with the scent of winter leaves. I've rented a small house, probably a converted barn, not far from the Musée Soulages in Rodez in south-west France. It sits on the edge of scrubby fields which give way to a swathe of deserted forest that stretches for miles, deep inland on a limestone plateau. During the day, an occasional farm vehicle stutters passed. There are no deliveries, no hikers; I haven't seen the postman. Now, at night, this is a place where darkness and silence fuse.

It's a month or so since Dad has been discharged from hospital, and we've come to this place together, renting a house for a few weeks. Just before we left, he had a consultation with what is kindly called The Memory Clinic. He ran through a series of cognitive tests with the psychiatrist, all of which confirmed the dementia diagnosis. Now we've come to a place of shared memories. We've had holidays together in this area; for a while I lived in the region, and dad would visit. I have photos of us climbing castle walls, sitting by rivers, playing table tennis under a seamless blue sky. Tonight, I've lit the fire in the house and left dad with a glass of wine and the quiet crackle of oak logs and dry beams. He was half asleep, or pretending to be, sitting in the pool of light from the table lamp.

It's the first time I've been out alone since we arrived several days ago, the first time I've left him. I thought he might have come with me, but he wasn't interested. He used to love the wilderness of the huge woods here, the skulk of deer, the bats, the quick pale slice of barn owl through the dark, but now I think all this scares him. He doesn't say so, but he might not know how. His language has skewed. Words run from him, skipping in and out of the dark like kids playing games. When he grabs at them, he falls. So he just says he's too tired to come and shrinks into the chair.

His anxiety has rubbed off on me. I feel edgy. I'm on a narrow dirt lane and can just about follow the uneven fence that edges the fields in the dark; glimpses of wire, the sense of a boundary. But then the track turns towards the dry stony woods of stunted oak, and the trees close in. Above, there are many stars and satellites, the elegant sparkling sweep of the Milky Way but here below, the dark shifts and pleats like a Soulages painting. I try to pick out the paler stony surface of the track against the thicker blacks beyond, but this turns out to be more difficult than you might think. Stable ground proves treacherous. I'm disorientated, wary. And as I make my way up an incline to the crunch of gravel and dry acorns, I begin to hear an odd sound, bodiless, clear and sharp but not quite physical. It comes at me from ahead and then from behind, from one side and then from the other. It seems close, but I can't judge accurately because it echoes, as though ringing on stone. In the dark, it's the rhythmic hollow clunk of phantoms and ghouls.

I turn tail. It takes me perhaps fifteen minutes to return to the house and then I retrace my route by car, still trembling. The headlights show a hay barn, a fenced track into the woods, the overhang of the trees, last autumn's leaves banked in ditches … and the donkeys. Donkeys! Five or six donkeys have broken loose from their paddock and are grazing happily in the dark, nudging each other along the verge, languid, unconcerned. *Clip-clop*. Hooves on packed dirt. The ordinary recomposed. A lesson in the power of night.

'When men in darknesse goe, they see a bush but take it for a theefe,' warns a seventeenth-century proverb. The Zuni peoples in Western New Mexico have a similar saying: 'After dark, all cats are leopards.' Darkness distorts, confuses and conceals; it transforms the mundane into the terrifying. No wonder dad is afraid. In the car headlights, my donkeys are just, well, donkeys. But stranded in the dark with just the sound of them, my mind spits panicky messages about all the things that *might* be there, all the possible dangers, all the terrors. We are scared of the other in its many forms, and it's in the darkness that this other often resides.

In the Middle Ages, as towns started to grow across Europe, rules were quickly introduced to try to keep such monsters under control. Almost everywhere, the carrying of arms after dark was prohibited, but so too was wearing anything that might act as a disguise, including hooded cloaks and broad hats, visors and masks. It was important to recognise your neighbours, and any strangers; to discern the bushes from the thieves. In many places, anyone walking the streets in the dark was legally required to carry a light, not only to help them pick their way through the narrow, mucky lanes but also so that their clothing and other marks of rank could be determined and their identity confirmed. Penalties were particularly severe: in fourteenth-century Paris, walking in the dark without a light carried a fine of ten *sous* which was enough to buy sixty loaves of bread, a massive amount for any ordinary citizen.

On dark nights, in unlit towns and villages, there was always the possibility that outsiders might sneak in to wreak harm or havoc; perhaps even more disconcertingly, without the evident trappings of status, social order might be threatened by citizens using the dark to challenge their place in the community hierarchy. It's no coincidence that it's at night that many of Shakespeare's characters shuffle through the comedies of mistaken identity. In the dark forest outside Athens, for example – and with the help of a magic potion – the four lovers of *A Midsummer Night's Dream* confuse themselves and the audience, while the boundaries of social power between human and fairy, noble and 'rude mechanical' become blotted in the dark. Chaucer, too, plays with the possibilities of darkness: 'dark was the night as pitch, aye dark as coal' when the chaos of misidentified body parts takes place at the end of *The Miller's Tale*.

Centuries later, the same potential for mishap and misunderstanding is gleefully exploited in Peter Shaffer's one-act play, *Black Comedy*. First performed in 1965, the action begins when a fuse blows in an apartment block and the lights go out, throwing the characters into complete darkness – although for the audience, light and dark is reversed. The darkness allows for silliness, for slipping off daytime constraints. It enables a jumbling of identity, and the discomforting laughter of confusion. But for all the farce, such tales also propose inherent threat – of losing ourselves or the one we love, of chaos, or uncontrollable consequences. For a while darkness might be comic. It might enable us to act out of character. But ultimately it frightens us. The chilling psychological encroachment that Conrad manipulates so skilfully in *Heart of Darkness* is effective because we are primed to look out for the darkness lurking just the other side of the ordinary.

In 1897, Stanley Hall, a friend of both Freud and Jung, wrote a study of fear that highlighted how easily the dark seeps into our minds to warp our judgment. He traced the beginnings of this process back to early childhood when our imaginations take off and 'impossible' monsters seem all too real. 'It is in those drifting automatic states so favoured by darkness,' he wrote, that children glimpse 'impossible monsters, ghosts etc'. But even in adulthood, Hall suggested, the dark continues to have a powerful effect. Night, he explained, was a cauldron of darkness in which all our terrors are born, 'the mother of fears', the dark returning us to 'ignorance' and a jittery state of terror which 'rules our nerves and pulses in the dark, despite our better knowledge'. An article published around the same time in the *Lancet,* with the evocative title 'Notes on Distressing Awakenings', also emphasised the capacity of the dark to muddle rational thought, throwing us back on the fearful instincts of childhood: 'The darkness of the night reduces many ... to the mental level of a child,' it noted, adding that, 'Thoughts grow far more extravagant in darkness.'

There is some basis, perhaps, for such 'extravagant' fears. In the dark, nothing is as it seems and nowhere is safe. Accidents and crime are both more likely at night, even now. In the past, when many communities lived in unlit villages or towns, the chances of falling into a stream or over a cliff, or of becoming the victim of burglary or a beating were high. The protection of civic authorities during the hours of darkness was slight at best – there were few night patrols, for example, even in large towns – so in an attempt to discourage nocturnal crime, curfews were widely enforced and many courts introduced stiffer penalties for offences committed at night. In the Middle Ages, burglary that took place during the hours of darkness was subject to the death penalty, a sentence that was vigorously pursued: in 1500s Middlesex, over four fifths of burglars were sentenced to hang. For many of our forebearers, darkness suggested a very real threat to home, possessions and life.

Are we right then, at some level, to be afraid of the dark? Should we take notice when the child inside us screams as the lights go out?

Since Hall's work at the end of the nineteenth century, more recent research has suggested that night-time fears are a normal part of a child's progress towards a full understanding of its place in the world. An instinctive fear of the dark emerges in nearly all children, no matter what their culture or upbringing, around two years of age. They begin to be aware of anxieties which have, until that point, been latent. This fear intensifies until around the age of five: in the dark, their disquiet taking terrible shape, the boundaries between fantasy and reality become blurred without the clarity of daylight. The 'conceptual cognition' of an ordinary wakeful state, when we're able to think rationally and use our experience to problem-solve, gives way to a 'primordial cognition' which is a more irrational and free-flowing state of thought, dominated by concrete images which appear all-too real (not unlike a vivid dream state). As we drift more towards this 'primordial' way of thinking, our fears can all too easily metamorphose into real monsters, or overwhelm us with a sense of getting lost in the darkness. By the impressionable age of seven or eight, most children report at least mild fear of the dark, articulating anxiety about ghosts and scary dreams but, most of all, talking of their consuming dread of being alone.

Early education manuals attempted to overcome this inbred fear by suggesting ways of gradually acclimatising children to life with the lights out. *Dialogues on the Passions, Habits and Affections Peculiar to Children* (1748), for example, recommended exposing children little by little to the dark by making up a series of errands for them to run alone at night. For those preferring to keep their children safely indoors with the family, the Swiss philosopher Jean-Jacques Rousseau recommended 'night games'. In his essay, *Emile, or On Education* (1762), he explained that this would help them take a matter-of-fact approach to possible night-time fears as they grew older, dampening the alarming instincts of childhood imagination:

> Accustomed to having a good footing in darkness, practised at handling with ease all surrounding bodies, his feet and hands will lead him without difficulty in the deepest darkness. His imagination, full of the nocturnal games of his youth, will be loath to turn to frightening objects. If he believes he hears bursts of laughter, instead of belonging to sprites they will be those of his old comrades. If an assemblage appears, it will not be for him the witches' sabbath but his governor's room.

But for all such efforts to eliminate, or at least suppress, our innate fear of the dark, it never goes away. Is this because we cling to it? Because it allows us to reinhabit the intense emotional states of childhood, and to experience the tingle of an unseen terror? 'The oldest and strongest emotion of mankind is fear, and the oldest and strongest kind of fear is fear of the unknown,' wrote the American horror writer H. P. Lovecraft in 1927. 'Children will always be afraid of the dark, and men

with minds sensitive to hereditary impulse will always tremble.' Wouldn't those of us of sensitive minds be sorry to lose
this 'hereditary impulse'? Think about the classics of Gothic
literature. They happily restore us to the monsters, ghouls
and shapeshifters of our childhood imaginations, delighting
in the dark precisely because of the confusion and fear it suggests. Towards the beginning of Mary Shelley's novel (1818),
Frankenstein recounts his journey back to his childhood home
near Geneva in the Alps. At first, he weeps with joy to be back.
The mountain summits, he says, are 'clear', both the sky and
lake 'blue and placid'. But the mood quickly changes: 'grief and
fear overcame me. Night also closed around.' He is plunged into
the dark, and with it, into a state of foreboding. From there it's
a short step to his monstrous fate. Edgar Allan Poe's creepy
poetic narrative, 'The Raven' (1845), plunges us straight into
darkness, beginning 'once upon a midnight dreary', a lonely,
dark state of 'fantastic terrors'. When the narrator opens the
door to find out who has come knocking at such an hour on a
bleak December night, he discovers 'darkness there and nothing
more'. Darkness the precursor to, and manifestation of, fear.
When he finally sees the raven, he describes it as a bird of the
dark, 'grim, ungainly, ghastly, gaunt', come 'from the Nightly
shore', a harbinger of death. Similarly, in *The Strange Case of Dr
Jekyll and Mr Hyde* (1886), Robert Louis Stevenson thrusts us into
a Victorian London of fogs and shifting darkness:

> here it would be dark like the back-end of evening; and there
> would be a glow of a rich, lurid brown, like the light of some
> strange conflagration; and here, for a moment, the fog would
> be quite broken up, and a haggard shaft of daylight would
> glance in between the swirling wreaths. The dismal quarter

of Soho seen under these changing glimpses, with its muddy ways, and slatternly passengers, and its lamps, which had never been extinguished or had been kindled afresh to combat this mournful reinvasion of darkness, seemed … like a district of some city in a nightmare.

The dark renders familiar streets strange and frightening, transforming them into a dreamscape in which the monstrous Mr Hyde emerges from the shadows.

Even Victorian novels that we don't necessarily consider part of the Gothic genre are full of a lurking darkness, but here we sense the looming threat not in ravens and dismal Soho streets but as something internalised, our fears turned to what might be skulking inside us, a 'psycho dark' that threatens the hero from within as well as without. In Charles Dickens' *Great Expectations* (1860–1), Pip discovers a 'vivid and broad impression of the identity of things' on a raw late afternoon as the dark draws in. He explores, for the first time, the landscape that is to become a defining feature of his life and of the novel, and with it himself and his position, seen through a 'dark flat wilderness … the low leaden line beyond was the river; and … the distant savage lair from which the wind was rushing was the sea'. In Charlotte Brontë's *Jane Eyre* (1847), lonely and abandoned orphan Jane is locked in a gloomy room on her first night in her aunt's austere house. Looking into a mirror, she catches sight of herself transformed into a ghoul in the darkness, a 'strange little figure' which she fails to recognise, this other self distinguished by 'glittering eyes of fear':

Returning, I had to cross before the looking-glass; my fascinated glance involuntarily explored the depth it revealed. All looked colder and darker in that visionary hollow than in

43

reality: and the strange little figure there gazing at me, with a white face and arms specking the gloom, and glittering eyes of fear moving where all else was still, had the effect of a real spirit.

We see a similar rebound between inner and outer in Maggie Tulliver, George Eliot's headstrong heroine in *The Mill on the Floss* (1860). Maggie is referred to time and again as 'dark', a girl of 'dark' hair and eyes, but also of dark spirit, of transgression, of 'eager and passionate longings', identified with the restlessness of the River Floss, 'the dark river that flowed and moaned like an unresting sorrow' and which eventually bursts its banks and casts the entire surroundings into darkness. At the end of the novel, as Maggie takes to the water and battles the flood in her little boat, she sees the village, St Ogg's, as a 'dark mass' but still she heads out on her reckless rescue mission because she imagines 'long-loved faces looking for help into the darkness and finding none.'

$\star^+_+\star \; \mathbb{C} \star^+_+\star$

Darkness and fear. Darkness as fear. The two are so entwined that we struggle to untangle them. When we look into the dark, we're afraid of what we might see. Or of what, like Maggie's long-loved faces, we might have to confront – the truth of our aloneness; the realisation that no one is coming to help us.

When things are about to go wrong in a novel or a film, when fears are about to be realised, we've come to expect the dark to close in. Towards the end of Harper Lee's *To Kill a Mockingbird*, for example, the young Jem and Scout go to a Hallowe'en party. It seems like a happy event; Scout is dressed, comically, as a ham hock. We're told, however, at the opening of the scene that 'there is no moon', and 'it was pitch black'. Just as Jem says, 'Didn't know it was this dark. Didn't look like it'd be this dark earlier in the evening', the pair are jumped by one of their schoolfriends, Cecil. The mock attack is laughed off, but it sets up a sense of unease nonetheless, and when the pageant is over, Lee reminds the reader several times of the darkness surrounding the children: 'It was still black dark', Scout notes, the schoolyard, again, 'black'. As they begin to walk home, the lights go off in the auditorium and they have to pick their way through familiar streets, Scout further hampered by not being able to see out of her costume. It's 'in the dark', that they become aware of footsteps following them – creating a tremor of fear for both the children and us. From out of the black night, a much more serious attacker emerges.

So moments of such horror – and also full-blown horror stories – discomfort but also entertain us. We sit on the edge of our seats while Jem and Scout pick their way home in the dark, feeling their fear but enjoying the tension of the night-time walk. In the hiding places around the school, we know there's something waiting for them. In this case, it's *someone* waiting: the threat is human. But often horror stories offer a superhuman or supernatural danger that doesn't respond to reason; a force that won't be repelled by the usual defences – something worthy of our fears. We can't usually shoot the monsters in a horror film, or entrap evil spirits with the force of the law. They are beyond normal means of control or defeat. We have to run away.

Worse still, the threat often refuses to show its face, lurking instead out of sight. Just as in the passage from *To Kill a Mockingbird,* the menace is sensed rather than seen. Darkness closes down the usual ways of working out what's going on and who's around, so instead the danger has to be inferred from the sound of breathing or footsteps or shrieks; from a trail of blood; from a shadow passing a window. In Stephen King's classic horror novel, *It* (1986), it's the repulsive smell of the predatory clown which sets up the terror:

> Smells of dirt and wet and long-gone vegetables would merge into one unmistakable ineluctable smell, the smell of the monster, the apotheosis of all monsters. It was the smell of something for which he had no name: the smell of It, crouched and lurking and ready to spring.

In *It*, the scariest of fiends, 'the apotheosis of all monsters' has no roar, no weapon, no name; 'crouched and lurking and ready to spring', it is invisible, intangible, no more (yet) than a bad smell. It is fear implied. This is how darkness works to tease us with horror, frustrating visual cues, disorientating and disarming us. When we're watching horror films, we respond quickly to unexpected moments of fear – what filmmakers often call jump-scares. MRI scans show how we react with the part of the brain responsible for processing emotion and making decisions, to elicit a rapid and appropriate response to sudden danger. Our brains are kick-started into action, immediately alert. But more powerful still is the response to slow dark scenes of suspense. As anxiety gradually increases, our senses become highly active, particularly the regions of the brain responsible for visual and auditory perception. We desperately scan for cues as to how to interpret the threat. Anticipating danger but not yet able to identify it, the brain searches restlessly, preoccupied, vigilant. But in the dark, we can't rely on what we can see. We can't read the signals. The suspense grows. We become more afraid. We try to catch small sounds and interpret them; we glimpse shapes and fragments and attempt to decipher them. We suspect the scent, the 'ineluctable smell', of monster.

The brain's tendency to fill gaps in perception with memories of things that have scared us in the past – to resurrect our latent fears – intensifies even more this satisfying tingle of the dark. Unable to see the villain in front of us, our imaginations cast in his place a villain we've come across before; our terrors suspend us out of time, eliding the then and now, the terror already over-come and the one still lying in wait. So children all over the world play a version of an invocation game where they stare at a mirror in an unlit room in an effort to glimpse demons or ghosts. The ritual has a variety of names, from Bloody Mary or Hell Mary to Hanako-san, but the 'trick' remains the same: the combination of nervousness and the dark lure the brain into conjuring the things that scare us. This is rarely something new. More often, it's a character we've come across before, an old fear that has long been lurking in the dark ready for its chance to pounce. Or more alarming still, it's not a ghoul or a murderer at all, but a more existential threat, a sense of physical danger, a horror of pain or disfigurement: in one study of the game, almost half the partic-ipants glimpsed a supernatural being, but almost three-quarters saw their own face deforming in the mirror.

This, then, is one reason why horror films love scenes that take place in the dark. It's no accident that the battle to sur-vive the night is a classic plot device. As the soundtrack ramps up the pressure and without accurate visual cues – as the film continues to insist that a killer or monster is poised to attack from the darkness – the brain makes scary leaps, automatically giving substance to the feeling of danger. It brings our fears to life. As far back as the 1890s, doctors were noting in the *Lancet* that the dark could incite 'a form of nocturnal insanity …

characterised by restlessness and excitement'. They found that people were subject to hallucinations or a 'disagreeable' sense that things around them were not where they should be – that a bed, for example, had shifted into a different position. Fortunately, all these symptoms of a strange nervous disease were guaranteed to 'completely disappear during the day', but our ability to shape the dark means horror writers only have to suggest the worst to us, and we'll do the rest. John Carpenter, the director of *Halloween*, claimed that the shadowy slayer in the classic 1978 film, Michael Myers, works so well precisely because he's largely unseen and unexplained, 'an *absence* of character' onto which we can project our demons, collective and personal. Eskil Vogt, the Norwegian film director, wrote the 2022 supernatural horror film, *The Innocents*. Working in Oslo during the summer months, he explained how difficult it was making a horror film without falling back on the potential of the dark: 'In Oslo, the sun doesn't set until 10pm and it only stays down for a few hours. So we didn't have the fear of the dark to play with, which is a really big deal in horror movies. It's such a primal thing.'

Instinctive. Primal. Irresistible. Simply being in the dark increases what is called the 'startle reflex' – an automatic protective response of the body to abrupt stimulation. Starting with the head, muscles contract rapidly, through the trunk and down to the knees. Hence the traditional idea of our knees knocking when we're afraid. The physical reaction we can have to horror scenes, then, from sweaty palms to a pounding heart, is automatically heightened simply by the presence of darkness. We are hard-wired to tremble when the lights go out.

★⁺₊★ ☾★⁺₊★

Bats can navigate in complete darkness. They don't rely at all on sight; even a blinded bat can navigate well at night. The eighteenth-century Swiss surgeon, Charles Jurine, suggested they relied on sound: he plugged the ears of bats to prove that they use sounds that are inaudible to humans to find their way, skimming so effortlessly in and out of the trees. This was only finally accepted as true in 1938, when the American zoologist Donald Griffin recorded ultrasonic bat calls for the first time.

Unlike bats, we're out of our element in the dark; we are highly visual animals. For us, darkness is a state of sensory deprivation rather than of specialist adaptation. And as torturers will attest, sensory deprivation – and the anticipation of it – can elicit strong fear responses and panic. Throwing prisoners into the dark has been a tried-and-tested method of intimidation and punishment for centuries. Peveril Castle, in the Peak District's Hope Valley, was built in the eleventh century above a stream that emerges from Peak Cavern, a huge limestone cave known as The Devil's Arse until the name was changed to avoid offending Queen Victoria when she visited in 1880. Tour guides gleefully tell stories of how prisoners were thrown down a chute into the cavern by the castle jailers, to be driven insane by the experience of wandering in the dark.

This, though, might be a case of tales-within-tales, dark-upon-dark: the terrifying stories were most likely invented by the rope-making troglodyte families who inhabited the cave system, manipulating our fear of the dark to encourage visitors to buy more candle tallow so as not to share a similar fate.

In 1689, John Locke wrote *An Essay Concerning Human Understanding* in which he suggested that darkness was not naturally feared by humans: he claimed fear was simply a learned response that parents passed on to their children through spooky stories. But in a counter essay published in 1757, Edmund Burke rebutted Locke's theory, suggesting instead that our fear of darkness is universal and innate: 'Surely it is more natural to think that darkness being originally an idea of terror, was chosen as a fit scene for such terrible representations, than that such representations have made darkness terrible… [it] is very hard to imagine, that the effect of an idea so universally terrible in all times, and in all countries, as darkness, could possibly have been owing to a set of idle stories.' As he describes it, the sensory deprivation of being in darkness renders all of us vulnerable:

> For in utter darkness, it is impossible to know in what degree of safety we stand; we are ignorant of the objects that surround us; we may every moment strike against some dangerous obstruction; we may fall down a precipice the first step we take; and if an enemy approach, we know not in what quarter to defend ourselves; in such a case strength is no sure protection; wisdom can only act by guess; the boldest are staggered, and he who would pray for nothing else towards his defence, is forced to pray for light.

The alarm of a precipice looming, fear of unknown obstructions, no means of adequate defence – Burke's summary of our helplessness in 'utter darkness' chimes with the reactions I witness in dad more and more often as his dementia takes firmer hold. He startles at small noises. His senses signal danger – the smell of a few spilled drops of soy sauce spark panic about a petrol leak; the red ignition light on the central heating boiler convinces him of approaching disaster, even after the engineer has been called many times. His brain has thrown him into an intense coiled state of worry that I can only edge close to by being in the dark. So I allow my walks to test me. I walk further, darker. The nameless, bodiless fear of what-might-be-out-there sets me on edge. I recognise the reactions I've read about in the brain studies and seen in dad: the body primed and anxious; an odd movement, a small sound, triggering terrors.

One particular morning, early, I'm crossing a high field above the flat lands of Cheshire before sunrise. On the horizon, distant cities glow. Above are one or two stars, a flimsy strand of morning grey low to the east. It's early spring, the ground damp and the air sharp with promise. The path, I know, slopes through the ploughed soil towards a sudden dip, then a lane and a couple of houses. But it's too dark to see the path; a few yards from the gate my foot crumples in a rabbit hole scraped into the sandy soil. I stumble and fall. When I pause to catch my breath, I have to make an effort to stand still. I feel all the hiding and sinister things around and behind me; my brain tells me to get a move on, to find a place lighter and safer.

But I hold my ground, alert, breathing hard, hands sore and cold from the fall, and I look up. I see what might be a star rise from the horizon. But it's not a star, of course; not even a shooting star moves at this kind of trajectory. Besides, what I'm watching glides too uniformly and steadily. And there are two lights now, one behind the other; two and then three, each the same distance from the one ahead and behind. Three and then four. And as I watch, a long line of lights emerges from my left, low down where the dark touches the crest of the hill. This line, this chain, curves across the open sky until the first disappears into the land at the other side while still above me many others string out one after the other, trailing through the dark like a necklace draped on cloth. I stand for some minutes, and still the lights trail uniformly, doggedly, across the sky above me; as one fades into the horizon to my right, another rises to my left. They go on and on.

I don't write science fiction. I don't generally accept conspiracy theories. I like to think I'm rational, mostly, and reasonably composed. But I've just fallen in the dark and my brain is spouting phantoms and looking up at this trace of lights across the sky, I feel my heart thump. I don't understand what I'm seeing. My brain can't make sense of it. I can't see enough to piece together a suitable justification. So I'm afraid. Something is coming for me, for us, out of the dark, and it is this overwhelming sense of threat that my brain and body respond to.

Fortunately, this is not, after all, an invasion. These lights are not alien vessels or high-tech weaponry. They are, of course, just satellites. Anyone accustomed to looking at the night sky will have seen satellites tracking across, these not-quite-stars confusing for a moment, and I find out later that what I've witnessed is evidence of Elon Musk's Starlink project. With final ambitions for 42,000 low-orbiting satellites in a new mega-constellation providing cheap broadband, the project has so far launched around 2,000 in batches over several years; sometimes these are visible. I've been scared by space junk. But what if I hadn't been able to understand or explain what I was seeing, rationalise it or look it up on the internet? What if I'd had to go on walking in the dark, frightened by these lights, completely certain that I was in some terrible imminent danger bearing down on me without reason? What if I'd been where dad is?

The new moon gives hardly any light but it slices the blackness. Darkness is no longer absolute. Clouds drift, edged in grey, smudged. Suddenly there are possibilities. While it remains invisible, it's nonetheless a traditional sign of hope and luck. If you take all your coins out of one pocket and transfer them to another when you see a new moon, things should go well – you should ensure good luck for the next month. If you turn over any silver, likewise, good fortune should be guaranteed.

This old symbol of new beginnings marks the end of the blackest nights. For many religions and cultures, this is a sacred moment. Darkness overcome; life restored; the cycle of existence rebooted. Our worst fears put aside. The Māori New Year, celebrated in June or July, begins with the new moon, associated with Whiro, the personified form of darkness, also presaging illness and death; following this dark start to the year, hope and life are restored. Sometimes the timing of the new moon – and the new month it initiates – is worked out by calculation: the molad, 'birth', of the Hebrew calendar, for example, may differ from the precise astronomical new moon because of the complexity of the calculations on which the calendar is based. Sometimes, only the evidence of the eye will do. The lunar Hijri calendar takes the first physical sighting of the crescent moon as the signal to begin a new month. I discover that there are people whose job it is to make these sightings: the head of Pakistan's Central Ruet-e-Hilal Committee, for example, aided by a hundred and fifty specialist observatories, announces the arrival of the new moon each month. This is a man who searches the skies for a glimpse of light, of change. This is a man who bears witness to the breaking of the dark, reasserting the order of time. But his position has not been without controversy. In the past, worshippers from different Muslim sects have challenged the primacy of his witness; as technology has improved, scholars have debated the precise moment at which it's possible to say the night is punctured. The moon can be as untrustworthy as the dark. It comes and goes behind clouds; it may be more visible in one country or time zone than another.

The Central Ruet-e-Hilal Committee's attempts to locate us in time and space by pinning us to the new moon can be subject to doubt and dispute.

But still the moon is our beacon, our anchor in the dark.

Theories of dark matter and dark energy suggest that these account for about 95 per cent of the universe. By these calculations, we exist in a speck of light. A tiny glowing aberration in the vast dark. The thought of this buckles when I try to get hold of it. Time and space and the dark all collide; the imagination leaps but does not land. I find it helpful to go back to some old reading, and in my efforts to grasp scale and substance, I re-encounter a tale told by Bede in the early Middle Ages, around 731. Before the technologies that gave us widespread light, this, of course, was a time of darkness: the passage of the moon and stars was a shared preoccupation; the lengthening and shortening of days is commonly recorded in manuscripts; the fires of hearths, ovens, beacons and torches were a matter of much importance. And in this largely unlit world, Bede writes of the passage of a sparrow from the dark to the dark, by way of a bright hall:

> It seems to me ... that this present life of men on earth, in comparison to the time that is unknown to us, [is] as if you were sitting at your dinner tables with your noblemen, warmed in the hall, and... one sparrow came from outside and quickly flew through the hall and it came in through one door and went out through the other. Lo! During the time that he was inside, he was not touched by the storm of the winter. But that is the blink of an eye and the least amount of time, but he immediately comes from winter into winter again.

Bede's parable was primarily a religious one. He was attempting to convert King Edwin of Northumbria to Christianity by illustrating how life without God was brief and solitary, nothing more than a flash of brightness. He was trying to find a metaphor for the unlandscaped blankness of a world without belief (and we'll look at the relationship between belief and the dark in a later chapter). But fifteen hundred years later, Bede's evocative story also helps me grasp the hypothesis of dark matter which proposes, at a basic level, that we are living in the light and warmth of the medieval hall while all around and beyond us is winter darkness.

According to the theory of dark matter, there is far more to the universe than can be seen; beyond our barn, in the dark, there is a whole load of everything. But *because* there is nothing to be seen, even by the most powerful of telescopes or the most distant of probes, the existence of this matter can only be implied – for now – by observation and calculation. It's a theory built on informed speculation. There needs to be something, the science insists, to explain the existence of many of the galaxies, to account for the ways in which they move and evolve, to suggest why they don't just tear apart as they rotate at speed. The explanation seems to be in the presence of large amounts of undetected material – dark matter – which acts like a kind of cosmic cement, holding the universe together. The existence of dark matter was first proposed in the 1930s by Swiss-born astronomer Fritz Zwicky, and is now so widely accepted that it's included in the standard model of cosmology that forms the basis for scientific understanding of the birth and evolution of the universe. Without dark matter, science can't explain how we got here. It exists, we think, because we can track the effect it has on other things – it interacts with known stars and planets, for example, via gravity – but it's difficult (so far, impossible) to prove because it doesn't absorb, emit or reflect electromagnetic radiation, such as light. Because it is dark.

Experiments at CERN (the European Council for Nuclear Research) are constantly trying to catch a glimpse of the particles which make up dark matter. The Large Hadron Collider, built in a 27-kilometre-wide tunnel near Geneva, collides beams of protons in an attempt to actually make dark matter, while telescopes are trained to track down the kicks and shakes created when it hits atomic nuclei in underground detectors. With this policy of 'making, breaking or shaking', the scientists at CERN are confident that they will discover a dark-matter particle within the next ten years. But such a discovery would by no means solve the puzzle. Thrusting us further into the dark, there is also dark energy. This is huge and ubiquitous, existing in both space and time and appearing to be associated with vacuum – with the vacancy of nothingness. Like dark matter, we only know about dark energy by watching other things: evenly distributed throughout the universe, it's been detected by measuring the effect it has on the rate by which the universe is expanding. More powerful than dark matter, and the more dominant of the two, dark energy is a repulsive force rather like an anti-gravity, but while its existence is generally accepted – and can be described using some convenient fudges in Einstein's Theory of Relativity – it remains what NASA calls 'a complete mystery'.

The scale implied by dark matter and dark energy is beyond my understanding. I sit one night in my garden and try to grasp it, unpeeling my mind from the here and now, but I discover it's impossible to get hold of this endless state of disembodied darkness. In the end, my thoughts settle again on Bede's sparrow as a way of comprehending how we flit momentarily though the enigma of the dark. Bede's tale is set in the depths of a

Northumbrian winter. There is snow on the ground outside the hall, he says, and a storm blowing. These details add to the evocative atmosphere of this short parable and allow us to picture the inhospitable world Bede conjures; they emphasise the dark beyond the comfort of the torches and fires. His choice of a winter setting gives the story intensity because instinctively we feel that darkness is deeper in winter. The light fades early; the sun rises late – darkness is with us for longer; it brings discomfort and cold, a sense of gloom. 'Melancholy were the sounds on a winter's night,' writes Virginia Woolf in her 1922 novel, *Jacob's Room*.

My walks in the new-moon phase have been winter walks. I've been seeking out places without roads and streetlights, setting off without a phone or torch to explore the dark as thoroughly as I dare. I've been trying to understand it. Dad's dementia has made darkness personal. The dark sits inside him like a small black bud. And watching my dad, the dark has become real to me too. I've begun to understand how our relationship with it is instinctive and visceral, how much it is a part of us. I've seen the way it blinds us in broad daylight. For the first time, I've faced up to the truths of Bede's sparrow: we are circumscribed by the dark, this vastness that we cannot see or feel or understand. We are each a moment of light, a match struck in a cavern, throwing shadows.

CHAPTER 2

# Waxing

It's a fine evening in spring and I'm walking a wide old path in the Pyrenees. The trees are in brilliant new leaf; the day has been grassy and green. The farm where I'm staying is propped below me on a narrow plateau. Its cluster of barns and walls and courtyards is sheltered by a dip in the land and a handful of tall beech trees. This path edges out from the back of the farm, over two huge flat stones which lie across a fast stream, and up a rocky slope. It's an ancient route from the valley to a high crossing point in the mountains, bounded by gnarly dry-stone walls and lined with more trees, hazel mostly, the lithe limbs sprouting from coppiced trunks. At this point, it rises gently although later it narrows and becomes steep, climbing up into the lingering snow.

Here in the mountains, sunset is never where you think it should be. Sunset always varies, of course, depending on location or even which way your view faces, but in the mountains, it becomes a particularly movable and uncertain experience, light and time and the incursion of the dark slipping from valley to valley, between ridge and summit. It's entirely dependent on the huddle of peaks. At the back of the farm, where morning comes early, the sun will disappear long before it sets for someone higher up in one of the lodges I can see on the opposite slope, facing wide open skies to the west; I'll be shuffling through the shadow of a mountain while they are still in full daylight. In such commanding terrain, sunset can douse you in a sudden gloopy shadow under one peak before a turn in the lane has it skimming away from the warmth of late sun. The colours of the fading day can feel like an intimate show, a reminder that time and space are arbitrary, and experience inescapably personal. Here in the mountains, dusk is negotiable.

For me, for this one path of many mountain paths, the sun has set. The light darkened quickly, without spectacle, and now there's a glassy greyish-blue sky with a hint of nicotine yellow. The land has bulked up with shadow. I don't intend to walk far but the evening is beautiful; I dawdle. It feels as though the day has slowed as the light has fallen. My steps slow, too; my breathing. There's a touch of magic here in this twilight suspended in this nook of mountain. Glow worms spangle against the walls; a wispy mist slinks low to the ground. In the clear sky above, a waxing moon falls, leaving a slight trail of a cooler pale light.

I wonder about the dusk, its familiar strangeness; its other-worldly everydayness. I consider my expectations. Spirituality? Epiphany? Surely, at least, an intensity of sentiment. But actually nothing much happens. It's a pleasant evening stroll. I make my way to a bend in the path, after which it becomes steep, and I walk back down to the farm. The chickens that have been scratching in the yard all day, chuntering their conversations back and forth from barn to barn, have gone to roost. That's about all. But looking back, I remember this evening walk very clearly, the textures of the darkening foliage, the smell of distance, the sense of possibility and of endings. I remember the glow worms and the wrap of translucent mist. I remember it as special. Was it?

What was it about the dying light that set this experience apart?

In my discovery of the dark, I begin thinking about the push-pull attraction of dusk. Darkness is more than an absolute. It's a thing of shades and stages, of evenings and mornings as well as midnights. The dark is implicit in a gloomy afternoon;

it's already there in that moment when the sun slips from the vertical. Lurking offstage in the sparkling red of a sunset – the pantomime villain eyeing up the flounces of the chorus line – we feel it as a knowledge, a presence, even as we turn our backs on it to glory in the spectacle of light. For dementia patients like dad, dusk is often a time of particular agitation, when the impulse to do something takes hold even if it's not clear what should be done. In the fading light, dad rattles door handles and tries window latches, rushes from room to room, unloads his clothes from wardrobes and drawers and lays them all out for packing. He shifts furniture and makes the bed many times. He's overwhelmed by the irresistible imperative to act, to move, to organise. In this short period between day and night, he can't bear the thought of things left undone or tasks unfinished. He worries about presents he hasn't bought and appointments he hasn't kept. Dusk is different for him now than any other time of day; it has its own shape and necessity.

But what is dusk then? A preparation? A premonition? You'll know the feel of dusk immediately, its distinctive smells and sounds, but it will probably also hang just out of reach, ungraspable. Unlike dad, you might not often take much notice of it. But what would the dark be without the dusk? Or is that the wrong question? Is it rather – what nature of darkness is the dusk?

Let's start with the disappearance of the sun.

The brilliant red-orange sunset of romance is just a play of the light. When a beam of sunlight strikes a molecule in the atmosphere, it spews wavelengths in all directions – an effect known as 'scattering'. This scattering occurs millions of times before the beam even reaches your eyes: while a sunset might look calm, the light is spilling and bouncing all over the place in a frenetic end-of-day dance. The two main components of the atmosphere, oxygen and nitrogen, have molecules that are a thousand times smaller than the incoming light. So these small molecules scatter the shortest wavelengths: blues and purples. This is why we perceive the daytime sky as blue. And it's most likely to look a startling blue towards noon when the sun is overhead, and light takes a shorter path to our eyes. But at sunset, the angle at which the sunlight enters the atmosphere is different; the light takes a longer path. This allows the blues, purples and greens to be scattered out before reaching the lower atmosphere – leaving that swish of red and orange, those fire-lit clouds that mark the end of the day.

This scattering of molecules is just a physical process, a phenomenon of nature. It 'means' nothing and connects to nothing. It's not intrinsically beautiful or moving; it doesn't signal anything other than the onset of the dark. It's not even essentially red: it's only our social structures, our language, which attach the word 'red' and its connotations to this sight in the sky. 'You have forgotten that no resemblance exists between primary qualities, the dull and senseless stuff out of which nature is really made and the secondary qualities with which you add a meaningless and arbitrary meaning to the senseless and meaningless hurrying of matter,' the French philosopher and anthropologist, Bruno Latour, reminds us. But even when we know this reality of physics, is a sunset really 'dull and senseless', nothing more than the 'senseless and meaningless hurrying of matter'? Say the word 'sunset', and a picture immediately springs to mind. But not only a picture: something beyond the pictorial, too, a memory, a sentiment, a sensation. We can *feel* a sunset. That brief photo opportunity when daylight blooms and then fades into dusk, holding us momentarily from the darkness, has become charged with significance, personal and cultural, an expression of what this world is, and what it is to us. We look for sunset and rate it – we know a 'good' one when we see one – seek it out on beaches and at tourist sites, wait for it, perhaps for hours; film classically has the hero riding towards it as the credits roll. This 'senseless and meaningless hurrying of matter' matters.

Across time and place, sunset has occupied artists, writers, travellers and lovers. Long before it proved such a hit on social media posts, letters home from exploration, exile or war often strained the writers' powers of description in their efforts to capture the sunset experience. The nineteenth-century poet, Gerard Manley Hopkins, wrote an extended correspondence to the journal *Nature* in December 1883, describing a run of autumn sunsets which followed the eruption of Krakatoa – an event that caused unusually colourful sunsets to be noted across the world. His accounts combine the precision of meteorological observation with rhapsodic efforts to describe a display that he never quite manages to pin down:

> The red was first noticed 45° above the horizon … By 4.45 the red had driven out the green, and, fusing with the remains of the orange, reached the horizon. By that time the east, which had a rose tinge, became of a duller red, compared to sand; according to my observation, the ground of the sky in the east was green or else tawny, and the crimson only in the clouds. A great sheet of heavy dark cloud, with a reefed or puckered make, drew off the west in the course of the pageant: the edge of this and the smaller pellets

of cloud that filed across the bright field of the sundown caught a livid green. At 5 the red in the west was fainter, at 5.20 it became notably rosier and livelier; but it was never of a pure rose.

Poems and novels, too, are scattered with sunsets, some described more effusively than others. Daphne du Maurier, for example, evokes a simple Cornish moment 'while the setting sun dappled the water with copper and crimson' in her historical novel *Frenchman's Creek* (1941). In *Little Women*, Louisa May Alcott is a touch more lyrical: 'The sun was low, and the heavens glowed with the splendour of an autumn sunset. Gold and purple clouds lay on the hilltops, and rising high into the ruddy light were silvery white peaks that shone like the airy spires of some Celestial City.' More flowery still is this passage from the classic American novel, Jack Kerouac's *On the Road* (1957), which pauses to notice a south-western sunset: 'Soon it got dusk, a grapy dusk, a purple dusk over tangerine groves and long melon fields; the sun the color of pressed grapes, slashed with burgundy red, the fields the color of love and Spanish mysteries.'

Sunset transforms. Clouds become shining 'airy spires', landmarks of a 'Celestial City', while ordinary fields infuse with tones of 'love' and 'Spanish mysteries'. It's little wonder we're drawn back, eager to glimpse this brief expression of grandeur and magic. But in this transformation, language tends to labour and buckle, grappling for metaphor; there's never quite the right word to match the experience. Painters at least, unlike writers, have the advantage of the visual, but we can see that they, too, struggle with the same problems of indefinability. In the seventeenth century, Dutch artist Aert van der Neer, for example, painted a subdued but romantic scene in his *Landscape at Sunset* (*c.* 1650) with pink and yellow clouds casting reflections onto a river. The horizon sits low in the picture; figures and animals are shaded in the growing dusk. It's the huge open sunset sky which is intended to draw the eye. But he must have felt he'd missed the mark; something was not right about this sweep of fading light because ten years later, he was still attempting to capture the effect of such skies draining into the dark, returning to similar pinkish tones but this time in a winter scene with the river frozen solid (and busy with skaters). His fascination, perhaps obsession, with adequately depicting the last tints of sombre light brought him back to changing sunset skies again and again: there are eight further evening studies in the collection of the National Gallery. Nor, of course, is he alone. Many other seventeenth-century European painters, with their focus on the landscape as an autonomous, almost living subject, were inspired by the colours and contrasts of evening. Michiel Sweerts, a Flemish painter born in Brussels in 1618, sets many of his characters in the half-light around sunset with glimpses of an evening sky visible tantalisingly in the distance. A similar technique was favoured by the Italian, Annibale Carracci: his painting, *Mary Magdalen in a Landscape* (1599) captures the sun going down behind distant hills and the sky darkening around the reflective (and semi-naked) Mary. Beyond Europe, too, sunset was a popular subject, with many artists experimenting with inventive materials and techniques in their efforts to recreate the elusive tones. Chinese studios often added flecks of gold to the paint to augment the effect of the dazzling colours. Similarly, artists of the Momoyama and Edo periods in Japan used gold leaf and minerals to suggest the intensity of the spectacle.

By the eighteenth century, the boom in travel meant that the search for evocative sunset landscapes was gathering pace, with the fashion for the picturesque and the sublime adding to the mystique. Joseph Wright of Derby painted a moody *Sunset on the Coast near Naples* (*c.* 1785); the Welsh artist Richard Wilson painted a number of coastal and river scenes in Europe, making use of the reflections of light off the water in an attempt to capture the distinctive tones. At Burnham in Norfolk, clergyman Thomas Kerrich, who was also the University Librarian at Cambridge, demonstrated the growing fascination with the subtle shifts of sunset light in his sketches of Norfolk skies. But it was with the taste for romance and nostalgia in Europe in the nineteenth century that sunset came fully into its own. J.M.W. Turner painted *Sunset*, Pissaro *Sunset at Eragny*, Van Gogh *Sunset at Montmajour*, Rousseau *The Forest in Winter at Sunset*, and *Sunset near Arbonne*, and Monet's works show the Houses of Parliament, the Seine, Venice, grain stacks at Giverny and his water lily pond all in the evening glow. The German painter and draughtsman, Caspar David Friedrich, faithfully observed the lilac twilights of northern Europe through the seasons. His studies of evening skies include some stunning depictions of brilliant sunsets but, more often, the works rely on the effect of a sweeping subdued evening sky, with delicate yellows, whites and greys drawing the viewer into the scenes. More recently, Maggi Hambling, although best known as a sculptor, has worked evocatively with colour to render disturbing, almost violent sunsets; David Hockney – an artist preoccupied with the brilliance of light – has chronicled the variety of evening skies visible from his Californian home, and in Europe. His picture *The Eiffel Tower in Paris* (2017), made on his iPad, shows the distinctive Paris skyline against a flat purplish-pink sunset backdrop.

What these centuries of literary and artistic effort have made clear is that sunset, and the dusk which follows, continues to beguile and evade, to inspire and frustrate, a handful of moments dissolving, reinventing, magic at one glance, modest the next, each new evening remaking the interplay of the gaudy, neon and jewelled as the sun goes down. The twentieth-century French anthropologist Claude Lévi-Strauss captured this sense of vexation as he watched a sunset from a ship just outside the port of Marseille in 1934: 'If I could find a language in which to perpetuate those appearances, at once so unstable and so resistant to description,' he said, 'if it were granted to me to be able to communicate to others the phases and sequences of a unique event which would never recur in the same terms, then ... I should in one go have discovered the deepest secrets of my profession.' Every evening the sun goes down, but every sunset is 'a unique event'. This herald of the dark is a paradox of repetition, a reminder that time passes newly even when it doesn't seem to pass at all.

While most of us, with Lévi-Strauss, have come to accept that sunset escapes us, unpredictable and irreducible, the idea that dusk could be a matter for individual interpretation caused some disquiet among scientific ranks of the past. In pre-industrial communities, the arrival of darkness was commonly heralded by some kind of sign – the ringing of bells or the beating of drums, even the blowing of horns from the ramparts of fortified cities. Gates were locked as villages and towns folded in on themselves for the night; the movement of dusk as the year turned was an accepted, although flexible, way of marking time. But as work (and life) became more complicated, and industrialisation demanded greater standardisation, such habits were increasingly deemed inadequate. By the end of the nineteenth century, the unreliable and subjective artistic observations and rural traditions were being formalised in a more prosaic attempt to define the exact anatomy of dusk. As early as 1817, mathematician Thomas Leybourn was distinguishing between two states which he termed 'astronomical' and 'civil' twilight – basically, the difference between twilight as dependent on the position of the sun, and as experienced by us on the ground. While this might have seemed an esoteric exercise, the French physicist Auguste Bravais pointed out the importance of clarity of definition: 'The length of twilight is an element useful to be known: by prolonging the day, it permits the continuance of labour.' At first glance, this reads like a manifesto for employers and landowners to squeeze extra work from hard-pressed labourers, ensuring no beam of daylight is wasted in the push for profit. And this was probably true in many cases. But at a time when much activity relied on natural light, even indoors, the ability to plan work reliably could be useful for many classes of worker: for farmers and sailors, those who needed light for sewing or weaving, those running a house or stables or shop.

The debate about what constituted dusk and how long it lasted was not easily concluded, however. Just as artists returned to twilight in an attempt to capture the elusive moment, so scientists pored over all the possible ways of defining those slippery end-of-day minutes. By the beginning of the twentieth century, there were at least five definitions of twilight, as well as an entirely new term, anti-twilight, which was the muted version of sunset which appears looking east at dusk rather than west: 'the pink or purplish glow in the eastern sky after sunset … a red arch defined upon the eastern sky around a darkish-blue space.' In 1989, The UK government, in the *Road Vehicle Lighting Regulations* (1989) and *The Official Highway Code*, set the 'beginning of the hours of darkness' as thirty minutes after sunset, after which there are rules for the use of car headlights to avoid dazzling other drivers. More subtly, astronomers have now settled on three distinct stages of twilight. The length of these stages varies with the time of year and your location but there is broad agreement on the terms.

For all three stages, the sun has gone down. The first, civil twilight, occurs when it has dropped to less than six degrees below the horizon. So, when it has just slipped out of sight. (If you want to measure six degrees of sky at the horizon, it's a little more than three fingers' width held at arm's length). At this point, enough natural daylight usually remains to do outdoor activities; the sky is darkening but only the brightest stars can be viewed with the naked eye. These are those moments we generally think of as dusk, when, on a summer walk, a lone bright star hangs above the horizon in a bluish sky. We know that dark is approaching but it feels a long way off. In the UK, civil twilight usually lasts about half an hour or so.

The following stage, nautical twilight, occurs when the sun is between six and twelve degrees below the horizon. While we can still make out what's around us outdoors, artificial light can be helpful: while you won't walk into a tree, you would need a strong source of light to safely cut it down. As the term suggests, this definition refers back to nautical traditions, and was coined to pinpoint the stage of dusk when most stars are visible and, in clear weather, the horizon can also be seen. When sailors relied on the stars for navigation, nautical twilight marked the moment when the skills of reading the dark were 'clocked in'. Like civil twilight, this stage usually lasts about half an hour, but in the far north of the UK, in Scotland, there is not enough darkness in summer for twilight to ever sink to the nautical stage: in Edinburgh, for example, there's no nautical twilight from early June to mid-July.

The darkest official twilight state is astronomical twilight, when the sun is between twelve and eighteen degrees below the horizon. During the period of the summer solstice in the British Isles, this doesn't occur: in Southampton, in the south, there's no astronomical twilight between the end of May and the middle of

July; at the far north, in Lerwick, astronomical twilight is absent for four months, from mid-April to mid-August. During the rest of the year, it's what most of us think of as night. While the atmosphere still scatters and refracts a small amount of sunlight, darkness is upon us.

Here in the west of Wales, the sun sets over the sea, sloping off into Cardigan Bay. It is, I suspect, civil twilight, on a clear, calm evening in March. I watch from a hill behind the coast path as the sky burns huge, the colours spreading and deepening. Everything around me bristles with the glow: tufts of coarse grass are tipped with orange and washed with gold; fenceposts shimmer; the greenish tones of the still-bare trees burnish to brass. I stand in the cold until there's only a faint glimmer of pink left in the sky; I watch the sun warp into the horizon and I go on watching, the salmon-orange tails of colour breaking and fading. Now, a slim smile of crescent moon appears over the sea; the light changes. This waxing crescent hangs for a while in the last smudge of yellowish-pink and then disappears the same way the sun did.

Describing the period between a new and a full moon, the waxing crescent grows larger each night until it reaches the first quarter, at which half of the moon's visible face is illuminated – the right half for those of us living in the northern hemisphere. As the moon grows larger still, we move into the waxing gibbous phase, in which the circle continues to fill until it's complete, but tonight there's the classic crescent of fairy tale. In this phase, the moon tracks the sun, rising and setting not in opposition but rather shortly afterwards, like a little brother trailing along in its wake.

And behind it, faintly, I can tonight see the outline of the rest of the moon. This is a phenomenon known as earthshine, when light from the earth reflects back and illuminates the entire disc, a state described memorably by Coleridge: 'I saw the new moon ... with the old moon in her arm.' (Although, for the pedantic, it's the young waxing moon, rather than the new moon, which the narrator actually witnesses).

When the sunset and the waxing moon combine in this way, it can feel special. But is this just learnt association? The afterglow of postcard idylls and romantic films? Maybe. But it might also be more than this. In many cultures, people have celebrated the close of day as a sacred experience, with sunset marking the transition from one state to another, and from one world to the next: Izumo, on the western coast of Japan, for example, has been a home of the gods since the very earliest Japanese writings. In the land of the rising sun, this place of sunset acts as a connection between different worlds; numerous temples and shrines have been built to worship and honour the gods; prayers and celebrations mark the vanishing of the sun into the sea. Alongside this sense of sunset as sacred, some accounts also point out that the colours of a good sunset are the same colours as a good fire: it's been proposed that sunsets return us to the warmth and security we've learnt, over thousands of years, to associate with the fireside. The wavelengths of light that cause the yellow and orange glow over the horizon are, the theory claims, recognised by the brain as especially comforting and attractive, so we're drawn to the distant sunset as in earlier centuries – and millennia – we might have been drawn to the promise of flames. Other studies point to the fact that the colours blue and orange – the constituents of the sunset sky – are at opposite ends of the light spectrum, an opposition our brains find particularly pleasing.

These ideas make some sense. It's long been appreciated that aesthetic pleasure, whether from nature or art, from colour or form, can result in positive mood and a general feeling of wellbeing. In the eighteenth century, the philosopher Immanuel Kant suggested that a lovely summer evening could evoke 'lofty sensations' which not only numb the nuisances of the everyday, but might turn our minds towards greater, more abstract, ideas: 'Temperaments with a feeling for the sublime are drawn gradually by the peace of a summer evening towards lofty sensations of friendship, eternity, indifference to the world,' he wrote in his *Observations on the Feeling of the Beautiful and Sublime* in 1764. His reflections on the effects of 'the peace of a summer evening' – while loaded with religious and moral overtones rooted in contemporary conversations – were borne out by later studies which showed that the experience of beauty is more than just a whimsical taste for a good sunset. Later research proved that the contemplation of something we regard as beautiful does, as Kant intuited, have a direct psychological effect, provoking a sense of calm and comfort. But it's not all in the mind: changes in mood and emotion are matched by physical reactions. Simply being outside to experience the beauty of a fine dusk has been shown to reduce blood pressure while increasing brain activity in the medial orbitofrontal cortex, which is the part responsible for pleasurable feelings of reward and satisfaction, the 'lofty sensations of friendship, eternity, indifference to the world' that Kant described two hundred and fifty years ago.

The deep-rooted, perhaps instinctive, appreciation of dusk and its 'lofty sensations' can be seen in the number of words we have to try to describe it. Some are bulky poetic terms that we rarely hear or use these days, and which sound ugly to modern ears, not quite matching the delicacy of the experience: 'crepuscular' from Latin and Middle English; 'stygian', originally relating to the shadowy world of the River Styx, and coming to mean the incursion of dark; 'tenebrous', from French and Latin; 'gloaming' and 'gloam' from the Old English word for twilight 'glōm': 'I saw pale kings and princes too … I saw their starved lips in the gloam', wrote Keats in his sombre 1819 poem, 'La Belle Dame Sans Merci'. More redolent, perhaps, are some of the dialect words: 'drop night' in Herefordshire, summoning the 'dropping' of damp pestilential air; 'dimmet', in Wiltshire, with obvious links to 'dim', and the long 'simmer dim' of northern Scotland; 'dimpsy' in use in south-west England; 'owl-leet' in north-west England and the more archaic, and consciously poetic, 'daylight's gate'. 'Cockshut' – when it's time to put the chickens away – sits in evocative opposition to morning 'cock crow'. 'Shutting in', a very common term, particularly in rural areas in the past, refers not only to the process of bringing in livestock for the night, but also closing the family into the house, pulling shutters and bolting doors to keep out the cold and protect the household from intruders. Irish Gaelic charts the movement of later afternoon into nightfall with four distinct terms, including *idirsholas*, meaning between light, and the marvellous *amhdhorchacht*, which describes the layered colours of a dusk sky and translates along the lines of 'raw darkness', in the sense of being uncooked, evoking the marbled layers of raw meat. In contemporary common English, by comparison, we're largely stuck with dusk and twilight, which we tend to use as synonyms (and I have done, in this book) but actually they're not quite the same thing. According to the Oxford English Dictionary, there's a subtle distinction: twilight, the earlier of the two words, from Middle English, is the subdued light we experience once the sun has slipped below the horizon; there is also a twilight in the morning, before the sun rises. Dusk is 'the darker stage of twilight', marking the end of the evening.

In the years following Immanuel Kant's meditations on a summer evening, the association of dusk with tranquillity and ease became commonplace. In her poem 'I know not how it falls on me', for example, even the tumultuous Emily Brontë describes a summer evening as a time of healing, 'hushed and lone', a moment of natural peace when the 'faint wind comes soothingly.' But sunset is not so simply a thing of warmth and contentment. As the sun dips, the pleasure of spectacle is laced with anticipation of loss; as the sky darkens and we turn away, we have to face those uncertain moments when the dark is renegotiated. In the transient experience that holds us between day and night, we're suspended on the cusp of darkness.

In his poem, 'An Evening Walk', written in 1793, William Wordsworth pauses to examine in detail the passage of day into night, and with it the ambivalent and changing emotions that are prompted by close observation of the states of dusk.

'How pleasant, as the sun declines, to view/ The spacious landscape change in form and hue!' he begins, tracing the ways in which the colours of the sunset transform ordinary objects: 'Even the white stems of birch, the cottage white/ Soften their glare before the mellow light.' But as night falls more quickly, hints of melancholy nudge through the initial enthusiasm. The poet-narrator notices a 'dusty cloud' of mist rising from the road ahead of him, gradually spreading across the scene. He describes it as 'a moving shroud', with all the overtones of death this embodies, before noticing that the only human figure in the landscape, a shepherd, is engulfed first by the light, described now as a dangerous force, as 'wreaths of fire', and then by the enclosing darkness. The poem moves from the celebratory language of its opening to a more wistful, even menacing closure as the shepherd's disappearance into the dark foreshadows his ultimate decline and death: 'The shepherd, all involved in wreaths of fire / Now shows a shadowy speck, and now is lost entire.'

The concept of 'melancholia' – that is, the psychological, physical and even moral implications of feeling dejected – was a familiar one with the Romantics but it wasn't a new preoccupation: it was the subject of extensive medical and popular debate from the time of Hippocrates through to the late nineteenth century. Understanding and interpretation of exactly what it means to be melancholy and what causes such emotions shifted back and forth during this time, but interest reached a peak during the sixteenth and seventeenth centuries when melancholia became something of a cultural and literary fashion choice. At this point, it was an emotion both desired and feared, a state of intense anguish associated with spirituality, madness, love, creativity, genius – and dusk fading into the dark. Writing in the sixteenth century, the Swedish clergyman scholar, Andreas Laurentius, noted, for example, that melancholia was a condition characterised by 'coldness and darkness'; Albrecht Dürer's large print, *Melencolia 1* (1514), shows the winged figure of Melancholy, seated head in hand as the sky behind darkens at dusk; the encyclopaedic, and highly influential, 900-page medical textbook, *The Anatomy of Melancholy* published in 1621 by Robert Burton, explained that melancholy could be induced at dusk by 'air ... dark, thick, foggy'. He asked, 'what will not a fearful man conceive in the dark?' noting that 'darkness is a great increaser of the humour.'

The idea that the uncertain moments in anticipation of darkness are naturally a time of melancholy, when thoughts turn towards sad recollections and death, proved hard to shift. A hundred and fifty years after Wordsworth's evening walk, a similarly disconcerting picture is evoked by Philip Larkin in his poem 'Going' (1955), which begins 'There is an evening coming in / across the fields.' Again, this opens with a sweep of rural landscape and at first, like Wordsworth's experience, the arrival of evening is shown as benign, even comforting: the word Larkin uses is 'silken.' But this is a poem which Larkin first intended to call 'Dying Day', and as the evening progresses, and dusk draws in, it becomes clear that 'it brings no comfort'; we begin to understand that we're reading about the final moments of life as well as the last light of day. As the landscape disappears into the darkness, the character in the poem is left disorientated and disturbed, fumbling with the weight of old age and finally losing any link with the world:

> Where has the tree gone, that locked
> Earth to the sky? What is under my hands,
> That I cannot feel?

> What loads my hands down?

I'm not alone, then, in feeling that an evening walk is a thing of mixed emotions, when the close of day alerts us to the fragility of the human condition; a moment for contemplation and even sorrow. It's something of a tradition. The twelfth-century Japanese writer, Kamo no Chōmei, reflected on the term *yūgen*, a concept of twilight which incorporates mystery and darkness, and which also refers to a style of poetry. For Kamo no Chōmei, an autumn dusk perfectly captured the instinctive but incomprehensible emotions of dusk, admiration for its beauty accompanied by sadness: 'although we cannot give any definite reason for it, we are somehow moved to tears.' We might also note that Thomas Gray's eighteenth-century 'Elegy Written in a Country Churchyard' (1751), which meditates so forcefully on death, the passage of time and the insignificance of human endeavour, begins at dusk as the cattle wind slowly home and the owl begins to hunt. As darkness seeps into the churchyard, the physical reality of the world recedes: 'Now fades the glimm'ring landscape on the sight,' Gray writes, and we move with the poet's thoughts towards the end of things.

Dusk, a thing of nearly dark. Of partings, the ends of days, the ephemerality of love and life. When I step back from the sea towards the hills behind me, I can already see the gathering of night: crows circle black above shadowed pines and beyond, the world reduces in the twilight. I think of my dad who now fully inhabits this space somewhere between light and dark, between being and not-being. His dementia has brought him here to a place of dusk.

Perhaps one reason we love sunsets so much is because they dazzle, briefly, the dark of night and death. They offer the momentary possibility that dark may not fall. But of course it does, as inevitable as our mortality. 'The sunset is infinitely good, but it is almost dusk', wrote the Chinese poet Li Shangyin of the ninth-century Tang dynasty in an evocative two-line poem. Earlier still, Roman writers also drew attention to the transience of beauty and pleasure: the Latin phrase 'et in Arcadia ego' is, unsurprisingly, not much bandied about these days but in earlier centuries it was recognised as a timely reminder of the brevity of life and its delights. Translating roughly as 'Even in Arcadia, there am I', it warns that even in the midst of happiness (Arcadia), death still lurks. As time passed, its powerful *memento mori* message proved a haunting subject for painters – who chose dusk as the appropriate time of day for rendering such a glimpse of death-in-life.

The phrase 'et in Arcadio ego' can be traced back to Virgil's long poem, 'The Eclogues', but it was in the early sixteenth century that the Italian poet Jacopo Sannazaro first created the influential idea of Arcadia as a beautiful rural idyll, a lost world of contented shepherds skipping through joyful days surrounded by music, poetry and love. A century later, this evocative land was taken as a subject by the Baroque painter, Giovanni Barbieri (also known as Guercino). His version of Et in Arcadia ego completed between 1618 and 1622, shows two young shepherds sitting under the trees contemplating a human skull in the forefront of the image, propped on a fragment of wall. The light is fading: the shepherds' shoulders and faces are lit but the rest of their bodies sink into the gloom of the undergrowth. The skull shines, highlighted, with the deep black pits of the eye sockets and mouth looming; the paleness of the light bouncing from it suggests a rising moon. Behind, the night sky encroaches. But it is not quite dark: this is still clearly twilight; streaks of grey and pale blue smear the horizon. For Barbieri, the natural state for musing on the passing of pleasure, beauty and ultimately life, is a state of dusk.

Several years later, the more celebrated French painter, Nicolas Poussin, revisited the 'et in Arcadia ego' story with his own painting of contemplative shepherds. In 1627, he produced a version with a tomb replacing the skull: three hearty men and their highly sexualised female companion have apparently stumbled across the overgrown grave half-buried beneath the trees, and are seen peering at the inscription. Poussin returned to the subject in 1637. This time, his female figure is fully dressed and standing with some authority over the three shepherds. The foreboding inscription is hardly visible on the blockish tomb. What remains constant in both versions is the twilight sky: the dying light of a yellow sunset in the first giving way to more subdued shadowy half-light in the second, fitting its greater austerity.

But why this age-old connection between dusk and melancholy? Why do our thoughts turn towards death as the day creeps away?

Such questions take us back to the wavelengths of different coloured light – to sunset – and then significantly further back, to the beginnings of life and ancient microscopic organisms called cyanobacteria.

First, the sunset: as the sun gets closer to the horizon and the shorter wavelengths are scattered in the atmosphere so, as we've already seen, the longer, redder wavelengths are more likely to reach us. This absence of blue light prompts our natural circadian rhythms to prepare for night – hence the repeated warnings not to boost unnatural exposure to blue light in the evening by watching too much telly or scrolling through emails before going to bed. These circadian rhythms are deeply embedded in all our body's processes, and are part of what makes us recognisable as humans. Which brings us to cyanobacteria, a lifeform known commonly as blue-green algae (although it is not actually, confusingly, algae). Cyanobacteria is prokaryotic, with simple cells that lack the tiny structures which perform specific functions in more complex organisms; they also lack distinct nuclei, which in other forms of life contain genetic material separate from the rest of the cell. Prokaryotes are the earliest life forms, and somewhere around two billion years ago, the cells from which humans are made evolved from these simple but hardy beginnings. So too did the circadian time-keeping system. Developing from cyanobacteria to humans, this linked our bodies intrinsically to a twenty-four-hour period, allowing us to anticipate, and respond to, environmental changes. In particular, it made us sensitive to sunrise and sunset. We became creatures of the light and dark.

Circadian rhythms are powerful and complex. They control hormone secretion, cellular function and gene expression, the process by which the information encoded in a gene is used to produce the protein molecules our bodies need. When circadian rhythms are disrupted, as a result of conditions such as depression, or neurogenerative diseases like Parkinson's, this can cause anxiety and mood change. Similar disruption also plays a part in 'sundowning' which is frequently experienced by dementia patients like dad, and which can make the hours around sunset a period of increased agitation and restlessness, perhaps even aggression. But even when our circadian rhythms are ticking over steadily, the dip in light at the end of the day still causes change, prompting clocks throughout the body to take note and prepare for sleep, eliciting shifts in temperature, blood pressure and hormones. Crucially for our understanding of the emotional implications of dusk, as the sun fades, our circadian rhythms signal a reduction in dopamine, with the body shifting instead towards production of melatonin, the hormone that controls our sleep cycle. Dopamine is a neurotransmitter that sends signals between cells. It's the molecule responsible for some of the most important elements of our 'humanness', from thinking and planning to emotional responses and motor skills. It's a complex and effective chemical made in the brain – in the substantia nigra, ventral tegmental area and hypothalamus – and then distributed throughout the body along a variety of neural pathways. Researchers continue to discover new ways in which dopamine influences our thoughts and actions but there's general agreement that it's responsible for many of the good feelings in the brain, for those warm sensations of reward and satisfaction. Many basic games and puzzles work by eliciting quick small shots of dopamine to boost us with a sense of achievement; many addictive drugs get us hooked by increasing dopamine release. It's probably no surprise, then, that as dopamine levels fall at dusk, the reverse can be true: we're more likely to experience feelings of melancholy, nostalgia and sadness. As dusk moves us towards the dark, our bodies are programmed to become more sensitive, even if only momentarily, to what the poet Samuel Taylor Coleridge described as 'grief without a pang, void, dark, and drear'.

Our circadian rhythms are so deep-seated and so influential that their disruption distresses our entire body. It can make us look and feel old before our time, influencing the processes associated with premature aging, while the systems controlled by the circadian cycle play a profound role in the most important physical functions, including cell proliferation, cell death, DNA repair and metabolic alteration. So when circadian rhythms are confused or disturbed over a significant period, this impacts the body's metabolic system, raising the risk of cancer and making us more likely to develop conditions such as diabetes, obesity or cardiovascular disease. Immune deficiency and high blood pressure – with the accompanying risk of stroke – have also been linked to circadian rhythm disturbance. Circadian disruptions can be caused by a range of factors, from long-term shift working and jetlag, to insomnia or disease. But dusk, too, is important. The loss of dark and its dusk precursor are just as likely to upset our body clocks as unsocial work patterns or sleep disorders.

Do you know when the sun has set? Can you mark the moment when dusk falls?

Streetlights, headlights, house lights, computers, televisions – all flash into life as the light fades; the eradication of the dark by powerful unnatural light sources is particularly evident at dusk.

There are those rare evenings when we're sitting out in a garden or a park, the day chilling as the sun slides away, a jacket needed, or socks, a blackbird singing, a few midges nagging – now we become distinctly, even mystically, aware of dusk. The dip of the light urges us to speak more quietly; our thoughts unroll into shadow. But most evenings aren't like this. Most evenings we're indoors, too busy to notice what's going on outside. We switch on a light when we need to; we look out of the window and the streetlights have come on; it's not clear whether it's dusk yet or not. It never will be.

Since the widespread adoption of electric light at the end of the nineteenth century, the boundaries between day and night have blurred. The distortion of time has allowed us to work longer hours, to punctuate the gloom of winter, to comfortably sew and read after dark. It's been a comfort and a distraction. We've gained control over the passage of the dark. But the quest for light has also cast us adrift, unanchoring us from the routines that put us together. If you send a woman to live in a cave without any light or heat stimuli, you'll find that she can maintain circadian rhythms, but over time, as she sits in the dark, these begin to shift. Lack of light deprives the body of the signals it needs to set its clock correctly; the complex tuning goes awry. So our cave-dweller begins to fall away from the rhythm of the twenty-four-hour cycle; her body loses its hold on day and night.

Light is the most important cue for maintaining perfect synchronicity with our environment. Until relatively recent years, these light cues were remarkably consistent. While there are always local variations caused by weather conditions or the lunar cycle, for any given latitude the regime of day, dusk and night has been consistent for extremely long periods of geological time, providing a reliable set of environmental signals for governing ecological and evolutionary processes. But this has changed. What has been a constant for millennia can no longer be taken for granted.

There was rudimentary streetlighting in major cities long before the Industrial Revolution. By the fifteenth century, there were a number of regulations in place in London, for example, requiring households on main thoroughfares to hang a lantern, at their own expense, on busy or particularly dark evenings, including Hallowe'en and Candlemas (2 February). York and Chester were among the cities to follow suit; Louis XI introduced a similar measure to Paris in 1461. But these rules only applied to the largest and busiest cities, and even here lanterns were often little more than candles, lit only for a few hours in the evening (wind and rain permitting) to allow people to find their way safely home at the end of the day. As late as the 1820s, many provincial centres, including Sheffield, Leicester and Norwich, reserved street lighting for particularly gloomy nights. Mostly the streets remained dark.

Now, though, we're losing our grip on the dark. Our exposure to artificial light is unprecedented in human history. Have you ever seen the magnificent sweep of the Milky Way? The clustered, textured, shift-shimmer arch of stars, dense and deep? If you've witnessed this, you're lucky: the spectacle is invisible to eighty per cent of the world's population (and ninety-nine per cent of the population of the USA).

Light hounds us through the day and into the night. Dusk is erased. We can't see the approach of the dark or feel it.

Such constant immersion in light changes the way our bodies work. Light leaches inside us, not only disrupting our circadian rhythms, as we've seen, but unsettling a range of other patterns

and processes. Hormones respond particularly sensitively. A state
of permanent illumination, without the repose of dusk, disturbs
the production of glucocorticoids from the adrenal glands which
are important in regulating stress responses: a disturbance in glu-
cocorticoid production is associated with a significant number of
mood disorders. A curtailed or eliminated dusk also means the
body fails to secrete the melatonin which we need to prepare for
sleep and which dictates sleep timing and quality. Studies have
shown that using a device like an eReader in the four hours before
bedtime can increase the time it takes to fall asleep and decrease
morning alertness; sleeping in low-level light can result in shal-
lower sleep and increased sleep disturbance, as well as markedly
decreased brain oscillations during sleep. These oscillations are
important because they're linked to cognition, so even relatively
low levels of night-time light could have detrimental effects on
cognitive function. Deprived of darkness, our brains don't work
as well.

So we can turn off the lights, shut down devices, black-out
the windows. We have some control over the amount of light
we allow in our homes and bedrooms – but outside, there's not
so much we can do. Light outside the home is enough to disrupt
sleep, no matter what measures are taken inside: those living
in areas with most outdoor lighting experience less quality and
quantity of sleep, as well as more daytime sleepiness; insomnia
and snoring are associated with areas of high light pollution. We've
grown so accustomed to a certain level of illumination even in
the middle of the night that we might not even notice it, but the
light intensity on an average urban street is estimated at 5-15 lux,
much less than daytime light, of course, but way beyond the 0.001
lux of a starry night and significantly more than a full moon on a
clear night, which generates 0.1-0.3 lux, or up to 1 lux in tropical

countries. This streetlighting, along with light seeping from shop windows and office blocks, housing estates and security lights, is more than enough to suspend us in a state where we can't move on towards the dark.

Dusk is important. It acts as a physical and emotional marker, and tempers the day, allowing for the evolution of complete darkness. But it's a tender thing, fragile and flowing. It needs to move, to shapeshift, to ebb. Light pollution creates a false non-dark, too hard and unreal, too monotonous to be dusk, a bogus blur of illumination rather than a gathering of the dark. Dusk defines the passage of light and time; permanent illumination denies both.

In this denial, we see the natural world beyond us also undone. Hatching turtles are disorientated; predators struggle to hunt. Camouflage ceases to conceal. Reproductive cycles fail. Artificial light extends the day causing animals to stay active for longer, exhausting themselves. The pall of light pollution triggers sleep disruption, changes to the brain, unexpected mating behaviour and an imbalance in species. For birds, flying to their roost as the light fades, from the dawn chorus to sworls of murmuring starlings, states of half-light at the transition from day to night are fundamental to avian rhythms, so when dusk is polluted, some of the most recognisable – and well-loved – aspects of behaviour are affected: in the glow of city lights, our birds sing less at the close of the day.

The natural world's relationship to, and reliance upon, dusk is so entirely integral that changes can affect the most unlikely and unassuming of creatures. Look at the humble sand hopper. These tiny, transparent, flea-like beasties have a subtle awareness of dusk and dark. They wait for the fall of light to navigate a nightly migration from burial spots in the sand to the open shore, where they feed on seaweed. But on beaches affected by nearby lights, where dusk is smudged and the signals marking light and dark are altered, sand hoppers become confused and disorientated. Some act as though day has burst on them unexpectedly, and hop-scurry back to the sea where they feel safe. Some fail to venture out in the first place. In either case, these foragers go without the food they need.

I don't suppose many of us love the sand hopper. They're ugly, alien little things that trigger from damp sand like too many worries. The trouble is, of course, that adaptations in the behaviour of one creature suggest much more widespread change. The same light pollution that confuses a sand hopper threatens an entire ecosystem, the many species living on beaches and rocky shores, and even on the seabed. Plants, mammals, insects and fish: almost three quarters of habitats on sea floors close to the affected sand hoppers are damaged by the leak of light from the land. Since seventy-five per cent of the world's megacities are now located in coastal regions, and coastal populations are forecast to double by 2060, it's easy to see how much of the world's marine habitat is at risk from the simple fact that dusk is disappearing.

Our awareness of such threats has grown, of course, in recent years, but those who have watched the skies and been alert to the melding twilight have been noting the effects of light pollution for years. Charles Dickens' *Bleak House*, published in 1852–3, opens with the memorable description of a foggy November late afternoon: 'Fog everywhere. Fog up the river, where it flows among the green aits and meadows; fog down the river, where it rolls defiled among the tiers of shipping and the waterside pollutions of a great (and dirty) city.' This is a dense, grubby smog, the dampness of autumn thickened by the smoke and grime pouring from factories and chimneys, the air full of 'flakes of soot … as big as full-grown snowflakes'. But, Dickens writes, it's not just a case of soot and poisonous emissions. Light, too, is corrupted. In this false night of fog, there's an artificial and unhealthy light:

> Gas looming through the fog in divers places in the streets, as much as the sun may, from the spongey fields, be seen to loom by the husbandman and ploughboy. Most of the shops lighted two hours before their time as the gas seems to know, for it has a haggard and unwilling look.

In this dingy world of fog and gas lighting, distinctions between day and night have disappeared, the natural dusk wiped out by the effects of pollution.

The Industrial Revolution had, among many other things, upended our relationship with light. As W. H. Auden wrote in his long poem, 'The Orators' (1932), dusk in the industrial towns and cities was no longer a decline into darkness but a time when 'the flare of foundries' became more visible. Some saw this as a positive indicator of growth: George Henry Thurston, a prominent Pittsburgh businessman and President of the Board of Trade, declared the 'shadow of smoke' over his city to be a clear 'report of progress'. But for others, the changes to our skies were a worry. Thirty years after Dickens' vision of a corrupt and distorted dusk in *Bleak House*, John Ruskin delivered a passionate lecture in which he drew on his observations of light and weather over decades to warn of the effects of pollution (and climate change), urging his listeners to take note of the way in which natural rhythms and cycles had been disrupted. He described the sunsets seen 'in entirely pure weather, above London smoke', which he'd observed frequently as a young man, and for the final time, he claimed, around 1876 – eight years before the lecture. This kind of 'old-fashioned sunset' was 'gold and vermillion ... deep scarlet, and purest rose, on purple grey', cloud and some smoke visible on the horizon but the spectacle unaffected. He goes on to explain how, more recently, evenings have been clouded and gloomy, the distinctions between day and night obscured and the glories of dusk diminished. Speaking after the Krakatoa eruption which, as we've seen, had affected weather patterns across the world, Ruskin speaks ominously (and only partly in metaphor) of the plague-wind and the plague-cloud which usher in 'their peculiar darkness'. These are not just, he claims, a result of the volcanic activity but of a more widespread and pervasive pollution resulting from the processes of industrialisation. The result is a 'thin, scraggy, filthy, mangy, miserable' dusk, ugly and unnatural, so sapped of energy that it 'can't turn the sun red'.

My dad's dusks tend to be 'mangy and miserable' these days. As the months pass, the sundowning phase of his day becomes more anxious and febrile. He's desperate to escape the house before darkness falls; he's goaded by the sense of time running away. We know to leave him alone – he gets angry quickly because we don't and can't share his anguish; he doesn't understand why we don't care about the dripping of the last light into darkness. Dusk is gathering, and it intimidates. This is not just the sadness or nostalgia of the evening. This is pain and panic. His day is ending; this baffles and frightens him.

Generally, I like the soft wrap of dusk. I welcome the cooling of the air, bold twilight bird song, the rustle of animals in hedges, a wind gathering, gusting. But I want to try to understand this other dusk, an unsettling, menacing thing of uncertainty and jump-scares. So I've come to a place I don't know. The unfamiliarity seems important. I want to feel disorientated, vulnerable. I want to feel the fading of the light as a loss. I've driven, randomly, a couple of hours from home, peeling from main roads to smaller ones, following those into lanes as the sun dips to the west. I've pulled the car onto a level piece of scrappy land adjacent to a ruined building. It looks like an old barn or stable, the original stone walls no more than shoulder high – lower in places – and the roof fallen in, a Kerplunk tangle of beams. There's a stony track leading from the side of the building, fields on either side and the edge of a small woodland just visible in the distance. It's not really a place to be afraid of. Still, perhaps its ordinariness will be part of the experiment. Dusk may transform the ordinary. Anyway, I wait ten minutes or so until twilight has begun to fall, lock up the car, and set off to walk.

The moon is high in the sky, having risen without me noticing during daylight hours. It's waxing gibbous, meaning it's moving towards full moon but isn't quite there yet. Gibbous comes from the Latin meaning humped or hunchback, bulging, and I can see the bulge of this moon clearly, like a bloated stomach. It gives plenty of soft light; the dusk is long. This stroll is inescapably pleasant. Birds sing. The track makes for easy walking, and I go at a good pace. An animal moves quickly across the far corner of one of the fields; a fox probably, perhaps a small deer. I don't see it clearly.

I come to the end of the field. The track narrows to a path, and forks. Either direction leads me into the trees where the dusk is denser. It doesn't matter which I choose. I have no destination in mind and even though I try not to, I'm enjoying myself.

I take the left-hand fork. The path skirts the edge of the woods, barely within the line of trees. It's getting darker now, but I can make out my way clearly enough, holly mostly along the path, the larger trees beyond.

Then a dog barks. A deep bark somewhere close. A big dog.

My heart jolts into my throat.

★⁺₊★ ☾★⁺₊★

Dusk is defined by most dictionaries as both a state between light and darkness, and also as a *tendency* to darkness. Dusk tending to darkness, moving towards it, nurturing it, growing it.

Is this what dementia patients sense at the end of the day? Is this what turns a distant dog bark into a threat? Dark rising, prowling, the tendency of dusk to leave us stranded?

Gerard Manley Hopkins, in his notes on the Krakatoa sunsets, noted how twilight could disturb and even disgust: the colours of certain sunsets, he said, were 'like inflamed flesh.' In *The Grapes of Wrath,* published on the eve of the Second World War, John Steinbeck evokes similarly visceral discomfort: 'A large drop of sun lingered on the horizon and then dripped over and was gone, and the sky was brilliant over the spot where it had gone, and a torn cloud, like a bloody rag, hung over the spot of its going. And dusk crept over the sky from the eastern horizon, and darkness crept over the land from the east.'

Steinbeck's is a violent dusk, torn and gory, a creeping thing that hangs over the dark. It's the dusk of folklore and fairy tale, a time when dangerous things emerge ready to feed on the night, the prospering of witches and devils and evil spirits who practice their cunning in darkness. There's a feeling of perturbation and physical danger. It's as dusk falls that the will-o'-the-wisp lures travellers into swamps and marshes. The nightjar, or whip-poor-will, will suck the milk from goats or

bring ill-fortune. These 'corpse birds', believed to be the manifestation of the souls of unbaptised children, rattle a portend
of death, with Aristotle and Pliny the Elder warning readers
to listen out for their otherworldly twilight song. Ancient
Romans also feared the strix, a witch-bird of dusk and the dark
that preyed on human flesh and blood, especially the entrails
of children. Shapeshifting Scottish Kelpies, who lure their victims into the water to drown them, sing out as the light fades;
it's at sunset that the supernatural jinns of Arabic and Islamic
cultures come to life, ravenously hungry and likely to cling to
you, causing illness or even death.

In Spanish folklore, dusk is when many *duende* emerge from
their hiding places to trick or terrorise humans. In an essay on these
*duende,* the early-twentieth-century Spanish playwright, Frederica
Garcia Lorca, drew attention to their diabolical nature: these creatures of the dusk were, for Lorca, creatures of death but also a state
of being, 'a force ... a struggle ... secret and shuddering', a dark
emotion which expresses itself in art, music and dance, especially
in the tradition of Flamenco. And the *duende* in all these forms
is drawn to dusk, to 'the edge, the wound ... [drawing] close to
places where forms fuse in a yearning beyond visible expression'.
The *duende* live in 'dark sounds, behind which in tender intimacy
exist volcanoes, ants, zephyrs, and the vast night pressing its waist
against the Milky Way'. In Lorca's powerful vision, dusk is not
a time of solace, nostalgia or even sadness: it's a portal to huge
primordial forces (and ants), mystery and danger, and a kind of
annihilating irresistible creativity.

Lorca's evocation of dusk as a thing of swirling, otherworldly
menace reminds me of Edvard Munch's famous 1893 painting, *The
Scream*. This shows a figure walking along a boardwalk beside a
fjord. It's twilight. There's a lurid red-orange sunset smearing the

sky behind. The figure's pale face is what we're drawn to – frozen in a scream, the mouth is wide open, the eyes staring; its hands clutch the sides of its head in horror. Colours churn around it; the blues of the water and the green edge of land seem to bend and move. But there's an ordinariness to the scene, too. Two shadowy figures further back on the boardwalk can be seen enjoying a calm and pleasant evening stroll – there's no apparent explanation for the powerful emotional response that we see in the horror-stricken face. It's an image of unexplained, unanchored fear. In a poem written onto the frame of his 1895 pastel version of the work, Munch explained the unaccountable existential terror of dusk that led to the painting: as he walked along a road with two friends, the sky, he remembered, 'turned blood-red/And I felt a wave of sadness'. As he watched, the landscape was transformed, 'Blood and Flaming tongues hovered'; held in the grip of this 'great Scream in Nature', Munch was 'tired to Death' and 'quaking with Angst'. From out of nowhere, the dusk had engulfed him and, as for Lorca, the blood-red sky drew on a deeper distress, a powerful force that shook the earth but remained 'beyond visible expression.'

Looking further back into this terrible twilight, we might pause to consider Homer's *Odyssey* and its portrait of Hades. This place of the dead, the underworld, is not some realm of misery buried deep below ground but is described instead as a state of perpetual dusk. Located west of the known world, where the sun sets continually towards night, Homer's Hades is an in-between state where the living and the dead might meet. Similarly, the wonderful *Epic of Gilgamesh*, a poem from ancient Mesopotamia, describes a key moment when the Uruk warrior and king Gilgamesh faces up to Humbaba, the monstrous guardian of the forest. The encounter, the poet tells us, takes place in the uncertain twilight: 'the world grows dark, the shadows have spread over it, now is the glimmer of dusk.' When Gilgamesh kills Humbaba, confusion ensues; the natural order of the forest becomes chaotic and the land unstable: 'the mountains were moved, and all the hills.'

In the ancient world of Gilgamesh, dusk brings disorder, the confusion of light disorientating, allowing even the physical world to shift and slide. Centuries later, a short story by H. H. Munro, *Dusk* (1912), finds Norman Gortsby sitting on a bench in Hyde Park as the day fades to a 'shadowy gloom'. Munro, better known by the pen name Saki, paints a dreary picture of twilit London which takes us back to the uneasy dusks of folklore and legend, a moment when hidden creatures emerge, and daylight hierarchies are upended. As in *Gilgamesh,* this is a time of uncertain shadow and precariousness; like Lorca's *duende*, those who dare not venture out in the full light of day now appear on the streets, claiming this liminal time and space for themselves: 'men and women, who had fought and lost, who hid their fallen fortunes and dead hopes as far as possible from the scrutiny of the curious, came forth in this hour of gloaming, when their shabby clothes and bowed shoulders and unhappy eyes might pass unnoticed, or, at any rate, unrecognised.' Similarly, it's in the troubled period before nightfall that strangers appear in G. K. Chesterton's weird and disturbing short novel, *The Man who was Thursday* (1908). As the story opens, we're welcomed into Saffron Park in London, a workaday, featureless suburb which, at dusk, becomes a place of dreamlike eeriness: 'attractive unreality fell upon it about nightfall.' This particular evening is made memorable by its 'strange sunset' which, rather like Munch's painting, further transforms the ordinary daytime streets so that they 'looked like the end of the world'. This sunset, the narrator tells us, is a thing of 'violent secrecy', and it's while the town is in the grip of such an apocalyptic close of day that the odd poet, Gabriel Syme, arrives, 'as if he had that moment fallen out of that impossible sky.' This is a book with the subtitle 'A nightmare', and it progresses in a nightmarish confusion of dusk, drawing us into the misleading shadows of a bizarre anarchy and a half-light of hostility and ambiguity.

Strangeness and impossible skies. Violent secrecies and the end of the world. In the enigma of dusk it feels as though unexplained things are possible, as though we're closer to the unknown. In his 'discovery of dusk' watching the circling of crows, the naturalist Mark Cocker describes the eeriness and dreamy unfamiliarity of a landscape at dusk. Just as the hills and mountains moved for Gilgamesh, so he notes that 'things become less fixed ... commonplace items are blurred'. In this state of distortion and obscurity, the ambivalence between day and night, reality and unreality, the natural world, Cocker suggests, reclaims 'its powers of mystery'. And with this mystery, comes strange manifestations: 'ghosts emerge with the shadows.'

Here we have it then. Ghosts. Is this where twilight leads us, to the in-between state bridging life and death, a realm of phantoms and ghouls? Out of the familiar and the comfortable. Ghosts have haunted this discussion of the dark already, slipping into texts by scientists and philosophers, hovering behind Agricola's examination of the state of mining or Stanley Hall's study of fear. Dad glimpses phantoms poised behind the furniture, inhabits a changed consciousness that allows his dead parents to slip from room to room ahead of him and the mundane niggles of old meetings to burst out new and urgent again in the faces of former colleagues. But it's in the equivocation of dusk rather than the certainty of dark that such ghosts really show themselves. Fudging the certainties of day, is dusk the crack between the worlds; a door to the unknown?

Let's go back to the Latin word *umbra*, meaning shade or shadow. A word that had come to be linked to the dusk, by the early fifteenth century it had also taken on a figurative meaning, which saw it used to refer to 'darkness' in a broader, more metaphysical sense: darkness of the soul alongside the falling of night. By the 1590s, *umbra* had evolved further, fully swaddling the idea of troubled souls and now being used to describe phantoms and ghosts, shadowy apparitions of the half-light which crossed the threshold between life and death. This idea seemed to stick; in time, we see the connection between dusk and ghosts embedded in many cultures. In Anglophone writing, T. S. Eliot suggests '... ghosts return/Gently at twilight'; the American writer and editor William Keepers Maxwell luxuriated in long summer dusks during a visit to Ireland in 1968, recording that eerie moments occurred when 'the shape of things' was distorted 'and the supernatural is not at all implausible'. In his poem, 'The Darkling Thrush' (1900), Thomas Hardy takes us into a countryside 'made desolate' by the setting of the sun, 'the weakening eye of day' and notices that the frost is 'spectre-grey', the land lonely and the people who 'haunt' it disappeared. Only the song of the thrush breaks the solemn atmosphere, and this song, Hardy says, is a thing beyond the terrestrial, a mystical – perhaps sacred – expression of nature and divinity beyond human comprehension. The work of the early-twentieth-century writer Walter de la Mare is saturated by both the shifting light of dusk and the half-glimpsed suspicion of supernatural phenomenon. In his poem, 'The Robin', de la Mare describes the dusk as 'ghost grey' – 'ghost grey the fall of night … lurking where shadows steal' – and in his unnerving short story, 'Crewe', the prosaic first-class waiting room in a busy railway station is transformed into a place of ghosts by 'a murky winter dusk'.

In the obscurity of twilight, some spectacular ghosts stalk the fields and woods. The 'Roaring Bull of Bagbury', the subject of a nineteenth-century folktale, was the ghost of 'a very bad man', transformed after death into a rampaging bull. Charging through the village with great noise and commotion as day turned to night, he would 'roar till the boards and the shutters and the tiles would fly off the building, and it was impossible for anyone to live near him'. On Souther Fell, in the Lake District, a line of marching troops, calvary and carriages was reported travelling along the summit ridge for hours during the long twilight of Midsummer in 1745, although not a trace was visible the following day. The beautiful figure of Marion La Bruyere, dressed in white, has been spotted haunting the grounds of Ludlow Castle at dusk since the twelfth century, occasionally leaping to her death (again) from the castle's Pendover tower and being said to let out bloodcurdling screams on the anniversary of her suicide.

More common than such showy ghoulish spectacle, however, is a hint of something strange made real as dusk falls, a fearful suggestion or a figure half-glimpsed at a crossroads, unexpected noises in attics or barns, the shimmer of air below the trees. These quieter ghosts grow in the telling and become a part of the land-scape: in the past, every villager would know places to be avoided as darkness approached. In his collection of local tales and super-stitions, published as *A Provincial Glossary* in 1787, Francis Grose described how this panoply of ghosts could easily get out of hand for 'persons of all ages'. The fear of meeting a ghost after dark, he noted, 'embittered the lives of a great number of persons …

shutting them out of their own houses, and deterring them from going from one village to another after sun-set'. He found that rooms where people had died were frequently left unused for generations, or would be nailed up after a suicide, for fear of the ghosts that bumped and screamed in the darkness. 'In short,' he concluded, 'there was scarcely a bye-lane or cross-way but had its ghost … Almost every ancient manor house was haunted by some one at least of its former masters or mistresses.' Nor did such hauntings disappear with the more pragmatic demands of modern life. Here's an account from a letter to *Country Life* magazine in February 1942 describing a walk in the war-time half-light:

> In Scotland last year while walking through an ancient forest with my husband, we took a shortcut through the wild glen and intended to walk down the bank of the Fillen to Crianlarich … As we entered, my husband remarked: 'I don't like this place, it's too old and dead.' I was about to reply that I felt it only peaceful, but I suddenly had the sensation of depression almost amounting to hopelessness. What I 'saw' was more a feeling as if all about me was snow, under a leaden sky, and behind me there were people and their eyes were without hope. My husband saw that I was oddly frightened and so we left for Crianlarich. We told them at the hotel that we'd felt spooky at one place in the forest. The late Mr Alistair Stewart said: 'Oh yes, that would be where a whole village was lost in the snow and they all starved to death.'

In such tricks of dusky light, the past leaches into the present, time becoming porous and communities circling around violent and troubled histories. In those minutes when day has ended but night has not yet come, our sense of when we exist becomes as

unclear as where – just as dad's dementia breaks down chron-
ologies, making all the moments of his life simultaneous. In the
dusk, figures breathe through the thin light from the past; we slide
through the long-ago shadows. In Ancient Rome, the *crepusculum*
of twilight gave way to *intempesta nox*, night without time.

This subverting of light and time has been used to impressive
effect by many writers, evoking the dislocation of dusk. We've
already seen how Dickens' *Bleak House* begins with the foggy
twilight of a late London afternoon; later in the novel, when the
facades of status are beginning to crumble and the wealth and
power of the Deadlock family is on the brink of destruction,
we see Sir Leicester Deadlock holed up in his room waiting for
his runaway wife's unlikely return. This is dusk again: the day,
Dickens notes, begins to 'decline' and 'the gloom augments'.
Having cast off her marriage and the respectability that went with
it, Lady Deadlock is lost to both the house and decent society.
She is between worlds – between riches and poverty, honour and
ruin – and in many ways a ghost. The well-meaning housekeeper
tries to rouse Sir Leicester from his despair, realising that he's
attempting to deceive time and its processes: 'he is striving to
uphold the fiction with himself that it is not growing late.' As he
tries to keep alive the belief that his wife will come home before
nightfall, he clings to the odd no-time state made possible by
twilight. In desperation, the anxious housekeeper recalls him to
the present, and to a more concrete sense of time marked by the
chime of the clocks:

> I must, for your own good, and my duty, take the freedom
> of begging and praying that you will not lie here in the lone
> darkness, watching and waiting, and dragging through time.
> Let me draw the curtains and light the candles, and make

things more comfortable about you. The church clocks will
strike the hour just the same, Sir Leicester, and the night will
pass just the same.

We see a similar suspension of time in one of Dickens' short
stories, fittingly called 'To Be Read at Dusk'. Set in an Alpine
convent in Switzerland, we're presented with both a 'stained'
sunset of 'reddened snow', and with the dead bodies of unfortu-
nate climbers stacked in a nearby shed, their natural decay halted
by the environment: 'the bodies of belated travellers ... slowly
wither away, knowing no corruption in that cold region.' In this
oppressive timeless atmosphere, where the decay of death is put
on hold, talk among the mountain guides and travellers turns to
ghosts, premonitions and hauntings: one of them tells a story
featuring the mysterious Signor Dellombra – *dell ombra*, of the
*umbra* – who spirits away a young wife having appeared to her
months earlier in a dream; another describes how a dying man
appears to his brother on the other side of town in the moments
approaching death. In this trilogy of stories, time is bent and
twisted, the future and the past out of joint, and the order of
things overturned.

For the twentieth-century Japanese writer, Jun'chirō Tanizaki,
the dusky light of old rooms is the perfect environment for the
ghostly, making it 'simple for spectres to appear'. In this 'suspension of ashen particles', he writes memorably, 'always something
seemed to be flickering and shimmering, a darkness that on
occasion held greater terrors than darkness out-of-doors. This
was the darkness in which ghosts and monsters were active'. Like
Dickens, he invokes eerie spaces in which time as well as place are
loosened, allowing for the 'ghosts and monsters' to take hold. He
recalls the experience of entering a dark room, and asks: 'Have

you never felt a sort of fear … a fear that in that room you might lose all consciousness of the passage of time, that untold years might pass and upon emerging you should find you had grown old and gray?'

Tanizaki's question is put to dramatic effect in Simon Armitage's poem, 'Evening' (2006). Here, a boy sets off for an evening stroll. In a sleight of hand, we, the readers, become this boy, on the cusp between childhood and adolescence: 'You're twelve. Thirteen at most,' we're told. You slip out from home, and promise 'not to be long, not to go far'. The evening is young, 'there is still time … the peak still lit by sun.' But as the poem progresses, the light fades and it becomes clear that this is about a more complex journey, not a stroll from the back door but the passage from life to death. As the character goes on walking, dusk falls: 'Evening overtakes you up the slope / Dusk walks its fingers up the knuckle of your spine.' By the end of the poem, just a few lines later, when the teenage boy returns home, time has passed; there's a wife and child waiting. The poem ends: 'You thought / it was early. How did it get so late?' The evening has slowed and sped; dusk mires us in a moment that has become years.

In the accelerated present of contemporary living where time is often collapsed and digitised, a projected thing of flat screens and simultaneity, dusk allows for a reclaiming of ourselves as historical beings hovering between past and future, feeling our way through the ripples like water boatmen on a pool. In the wake of Armitage's boy, we begin to see through the deception of time as evening overtakes us; in the diluted light and uneven shadows of dusk we become our own ghost. While a sunset buoys us, suspends us in wonder, the dark creeps upon us, more beautiful and terrible. We hardly see it coming.

CHAPTER 3

# Full Moon

Rising early in the night, shortly after sunset, is a hunter's moon. The first full moon of the autumn, it looms behind the trees low to the horizon, a rusty silvery-brown, the blotches and shadows of the surface clearly visible. It's the end of October, clear and cool. The trees are not yet bare but there are fallen leaves are already gathering in the hedges; there have been one or two frosty mornings. The previous weekend, the clocks turned back; now darkness falls early, stretching the night. But the full moon, reddish, ready, opposes the dark. We're poised between soft light and deep shadow.

Traditionally, the hunter's moon marked a closing in of the year: a moment when fires were lit and larders stocked to meet the hardships of winter. The name evolved because it rises during prime hunting season, when communities turned their attention from crops and cattle to the wild game that would provide meat through the cold months. At a time of year when mammals and birds were fattening up to see them through a leaner period, this was a good moment to shoot and trap; the light from a full moon made this activity easier.

A hunter's moon signalled a hunkering down and shutting up, months of low light and low temperatures. The intrusion of the dark.

Hunter's Moon. Harvest Moon. Plough Moon. Seed Moon. Barley Moon. In astrological terms, the occurrence of a full moon is a simple matter of alignment: it occurs when the sun and the moon are aligned on opposite sides of the earth so that

100 per cent of the face of the moon is illuminated by light from the sun. (This is always the same portion of the moon's surface, due to it being tidally locked to the Earth.) But the litany of evocative names which we've given to full moons suggests something more than a scientific phenomenon. It links full moons throughout the year to the tending of land and the rhythms of human activity, to cycles of growth and our relationship to the world around us, to patterns of light and dark. Many of today's names are thought to come from native American culture, later adopted by colonial Americans, but Celtic, Anglo-Saxon and Norse traditions have also had an influence, and by the early eighteenth century, many of the terms were in accepted use. Simple verses were often used to keep track of the different phases of the moon, and almanacs became increasingly popular. By the seventeenth century, these were widely available across Europe: by the 1660s, over 400,000 copies of almanacs were published each year in Britain alone, with about a third of families estimated to own a copy. In early America, the almanac represented the most popular publication after the Bible. Sowing and planting, pruning, cutting, digging and harvesting all, in this way, took place at their proper times according to the state of the moon. Some contemporary growers, too, still follow these old traditions and work by the moon phases: French organic and bio-dynamic shops and co-operatives routinely stock almanacs; Sekem, a biodynamic farm in Egypt, has been farming 35,000 hectares by the moon in the desert north-east of Cairo since 1979.

The naming of full moons is testament to their enduring attraction, bringing them alive and drawing them into the mythology of the dark. Even though they're common, recurring more or less monthly, we notice them – they seem out of the ordinary. Every time we glimpse one, it feels like a new experience. As a result, many cultures choose this time for celebrations and rituals: several significant Hindu festivals take place at full moons, and the full-moon day, called 'Purnima' in Sanskrit, promises prosperity and happiness; native American tribes take part in a three-day moon dance which is considered to heal and re-energise; the Moon Festival in Chinese traditions, celebrated around the autumn equinox, is one of the most important holidays in the calendar, the round shape of the full moon symbolising family harmony and reunion; in Japan, a 'moon viewing' takes place around the same time, with picnics enjoyed outside to celebrate a successful harvest; in Buddhist communities, the fourfold disciples of the Buddha – monks, nuns, lay men and lay women – come together at full moon for special observance; lantern festivals are commonly held around South East Asia to coincide with the full moon. Across time and place, the full moon stands out in the dark sky, a beautiful thing, sacred and celebratory.

But a full moon is also a thing of strangeness. Considered to hold sway over madness and menstrual cycles, dreams and monsters, it's long been suspected as a disturbing, even pernicious influence. The Roman embodiment of the moon, Luna, gave her name to 'lunacy' and 'lunatic'. The Latin *lunaticus* means moonstruck, and so directly connects the moon to madness. Aristotle, Hippocrates and Pliny the Elder were among those who suggested that both brain and body are susceptible to the harmful influences of the moon, most evident at full moon, causing violent behaviour, seizures and mental illness. By the Middle Ages these ideas had become so tenacious that stories emerged across Europe of men overcome by bestial tendencies at the full moon, and the link between mythological werewolves and the full moon was born. By the sixteenth century, tracking the moon had become an accepted part of medicine – the full moon was considered a period of danger, when fever might take hold – and by the eighteenth, it was possible to plead at trial for a lighter sentence if you could prove you were under the influence of a full moon when you committed a crime.

Numerous studies have been done to try to prove some of these links, but so far there's little proof that the full moon can, in fact, make us mad, prompt a rise in births, disturb our sleep or transform us into monsters. Is there any change in dad's dementia when the moon is full? I watch carefully. Online dementia forums are peppered with accounts of sufferers becoming more unsettled during a full moon or their language disintegrating; families report an increase in anxiety or wandering. But dad seems oblivious to any changes in the moon cycle, and research has failed to back up such anecdotes. Because of its effect on gravitational pull, a full moon can sometimes – usually in spring and autumn – cause very high or low tides, but there's little evidence for much else. The lingering belief in the malignant power of the full moon may partly be a result of cultural influence: in Mary Shelley's *Frankenstein*, for example, the moon weaves in and out of the story, and is at its brightest when the monster is at its most mad and evil. When the full horror of the situation is finally revealed to Victor Frankenstein, it's by the light of a full moon: 'Suddenly the broad disk of the moon arose, and shone full upon his ghastly and distorted shape, as he fled with more than mortal speed.' Our perception that a full moon has some kind of unnatural influence on our lives may also be due to an effect known as 'illusory correlation', when we're seduced into believing a relationship between two things by a kind of psychological sleight of hand. A full moon tends to stick in our minds, so when another significant event occurs at the same time – a birth or death, an illness or a bout of insomnia – we remember the moon and assume there's a connection between the two, when in fact none exists.

It's true that the moon can drive physical processes for some animals: many inhabitants of coral reefs, for example, exhibit lunar cycles of reproduction, with both fish and corals spawning during a particular moon phase; birds and insects appear to respond to changes in the moon, and particularly to a full moon. But the myths around a full moon may simply have evolved because it has traditionally given us extra light to do things which might otherwise prove tricky. It allows us, for a few nights a month, to inhabit the dark. We can go outside to walk, dance, make love and commit crimes without the need for artificial light. We can harvest through the night and make sure the crops are safely got in. We can hunt the creatures of evening. The phrase 'moonshine', referring to illegal alcohol, derives from the smuggling of spirits across the Channel to the Sussex and Kent coasts during a full moon. Thrust outside of usual routines by the extra light, we might become more aware of unusual behaviour, in either ourselves or other people. We might throw off inhibitions; we may well feel the thrill of climbing a hill in the moonlight when we would normally be in bed. The full moon's silvery blend of light-in-dark draws us to prayer and mysticism but also allows a disruption of night. It makes us reconsider our relationship with darkness.

The human eye can function from very dark to very bright levels of light. The photoreceptors in our eye, known as rods, can be sensitive enough to detect a single photon (the smallest unit of light), and can detect light sources more than a billion times weaker than the light we see outside on a sunny day. There are about 100 million light-sensitive pigments in a single rod, and there are about 120 million rods in the retina. This makes us extremely susceptible to changes in light. In addition, our eyes are constantly adapting to keep track of light levels and to deal with contrasts between light and dark. Think about the experience of stepping straight into a darkened room on a sunny day. You can't, for a minute, see anything. It might take a while before you can even make out basic shapes. This is a process known as 'dark adaptation', a term coined in 1865 when Hermann Aubert, a German physiologist, first began to explore how the eye recovers its sensitivity after exposure to bright lights. 'Dark adaptation' might seem to resolve in moments but in fact the process can last for up to an hour. This is because our eyes break light down into molecules of a protein called rhodopsin in order to convert it into nerve impulses which the rods recognise. In very bright light, rhodopsin is bleached as quickly as it can be made, which 'turns off' some of the rods in the eye. At this point, our vision is taken over by cones which are concentrated in the centre of the retina – a switch which forms the basis of something known as Duplicity Theory. There are far fewer cones than rods in the eye (about 6 million compared to the 120 million), and they don't function in low light. So when you come in to the dark, you have to wait for the eye to switch on a supply of rhodopsin molecules to reboost the rods. When we move the other way – stepping out of a dark cinema, for example, into the glare of a summer's afternoon – the opposite process is known as 'light adaptation'. We're dazzled by white light as both the rods and cones are bombarded, producing a flood of signals. The brain quickly switches from rods to cones to cope with the extra light.

The functioning of the rods at low light levels explains some of the particular qualities we associate with moonlight, which appears to play its own tricks on the eye. About 400,000 times fainter than direct sunlight, moonlight is not intrinsically eerie, special or romantic – it's nothing more than normal old sunlight reflected from the dusty surface of the moon. But it seems to have its own enigmatic qualities. We think of a moonlit landscape as distinctive, a place of muted, metallic tones not unlike a black-and-white photo. Colour is drained and contrasts smudged. Strong reds and oranges fade into shades of grey, while whites and silvers stand out and seem to shine. If you're away from artificial light and you stare at this landscape for long enough, it may even seem to take on a wash of blue. This is the effect perhaps most famously captured by the American artist James McNeill Whistler in his series of *Nocturnes* from the 1860s and '70s. Depicting the grey-blue transparency of night skies, often over water, these are veiled, mysterious scenes. The paintings can seem lonely and unearthly, with forms and figures obscured and the deep black of the night only barely held at bay by the brush of pewter light; we recognise the effect instantly as the natural abstraction of moonlight in darkness.

What Whistler evokes in his 'Nocturnes' is the 'blueshift' or 'Purkinje shift', named after the nineteenth-century scientist, Jan Evangelista Purkyně, who first drew attention to it in 1825. A Czech anatomist and physicist who was famous throughout the world in his time, Purkyně noticed that a red rose looked inky black in the faint light of the moon; he went on to explore how longer, red wavelengths of light appear darker under moonlight, while shorter, bluer wavelengths appear brighter. To help capture this effect, and make night-time scenes feel more natural, many filmmakers put a blue filter over the lens. But it's not just colour that can play tricks in moonlight. During a full moon, it should be bright enough to read; if you open a book, the page will seem perfectly clear. But for most of us, there's no way of making out the words, which blur and fade and may even slink away completely, vanishing into a blind spot. It's the cones and rods, again, which cause these unnerving effects. Cones allow us to see colour and fine detail (like words on a page), whereas, as we've seen, the rods take over in low light levels. So, when we switch to the rods of night vision, under the influence of moonlight, colour is drained into a uniform grey. And because the centre of the eye is dominated by cones – rods are absent from a central area of the retina called the fovea – this becomes a night-time blind spot. So letters can seem to disappear from the page, and astronomers often choose to view night-time spectacles from slightly off centre.

Moonlight, then, like the darkness it attends, is a thing of perception, a matter (literally) of how we see the world. When it washes out the landscape to a monotone, it draws our mind to the colour we experience under more usual daytime conditions; glowing somewhere between light and dark, it makes us conscious of both. Perhaps most striking of all, it sets up a powerful relationship: where is the moon without the sun?

When children draw a sky, often both the sun and moon are visible, facing off from different corners of the page. Our understanding of one owes a great deal to the other, the two celestial forces of light shining in both harmony and opposition. The huge pale disc of the full moon offers us a light source we can gaze on, an object of contemplation in contrast to the unapproachable fire of the sun. Moonlight is the mysterious, romantic rival to sunlight. 'Arise, fair sun, and kill the envious moon,' Romeo famously urges, as he woos Juliet in the balcony scene of Shakespeare's *Romeo and Juliet*, a play built on antagonisms. Similarly, in Samuel Taylor Coleridge's long poem, *The Rime of the Ancient Mariner*, the sun and the moon are positioned as competing influences on the mariner's journey, with the sun lighting the vastness of the natural world and summoning powerful physical forces of heat and drought (it's thirst that finally kills the sailors) while the softer influence of the moon allows for the mariner's penance and his spiritual awakening.

Those who chart astrological patterns, believe that anyone born at a time when the sun and moon are in opposition (that is, they rise and set at very different times, around a full moon) are likely to experience competing energies leading to internal struggle and perhaps a confrontational nature.

Many ancient civilisations worshipped the sun and the moon side by side, recognising them as complementary but also contradictory powers, each with its own unique influence on human lives: the Pyramid of the Sun in the ancient Mesoamerican city of Teotihuacan in Mexico, probably built around AD 200, sits alongside the slightly smaller Pyramid of the Moon; two temples built around AD 500 by the Moche communities, near Trujillo in Peru, are dedicated to the sun and the moon respectively, the temple of the moon being the more significant and sacred site. In many mythologies, the sun is represented as a male force, and the moon as a female one. To the Tiv tribe in Africa, the sun is a male deity and the moon is his daughter. The Aztecs worshipped Huitzilopochtli, the sun god, and the moon goddess Metztli, who feared the sun's fire. The Incas' Mama Quilla was considered the moon mother and wife of the male sun god Inti. In Greek mythology, Helios was an energetic sun god, who drove a chariot daily from east to west, and Artemis was the alluring feminine goddess of chastity and the moon.

Two forces of nature; two sides of human experience; two articulations of gender. But the contrasts set up between sun and moon are only made possible because of the dark. Both sun and moon hang poised in defiance of darkness. Whether sunlight or moonlight, the breaking of the dark has been at the heart of our beliefs for thousands of years. We've learned to express our sense of self in the fundamental juxtaposition of light with dark.

★⁺₊★ ☾★⁺₊★

Walking in cities at night is an experience of deep contrast, of brilliance counterpointed with blackness. Mostly, it will be an experience of light. But the dark lurks. Alleys, doorways and corners shimmer with darkness of a new quality, a deep dusty darkness held in retreat, caged, but primed to defy its irrelevance. In his notebook, published in the late 1940s, the writer W. Somerset Maugham captured the particular harshness of the city dark: 'In the country, the darkness of night is friendly and familiar,' he suggested, 'but in a city, with its blaze of lights, it is unnatural, hostile and menacing. It is like a monstrous vulture that hovers, biding its time.' The monstrous vulture is a powerful image which stays with me, and when I swap the 'friendly and familiar' night of rural Wales for the 'blaze' of London, I confront a new dark, 'unnatural and hostile', incongruous. I'm attending an event in Trafalgar Square. The façade of the National Gallery is flooded with light; Nelson's column is illuminated; the roads around the square shine with streetlamps and crossings. Usually, when I'm walking at night, it feels as though any light source is an imposition on the dark but here things are turned inside out. Now light is the natural state, simple, distinct, definite. It's the tangle of darkness which is out of place.

I walk from Trafalgar Square to my hotel, which is tucked in a side street close to Westminster Abbey. It's a walk of less than a mile; it takes about twenty minutes. I do it on two consecutive nights, around ten in the evening, and am surprised by how deserted the streets are. The first night I take a path that skirts the edge of St James' Park, and then nip through a flagged stone passage called Cockpit Steps between the houses on Birdcage Walk; the second night I walk along Whitehall, past the entrances to Horse Guards Parade and Downing Street. There are no shops or modern office blocks on either route; this is the solid heart of the old city. And it's a patchwork of light and dark. When you walk here at night you feel as though you're crossing continually from one state to another; it's like being caught in the flicker of an old film. The park is mostly dark; the pavement marks a kind of boundary, the other side of which the streetlights hum light. I walk along this seam, night on my right, my left arm swinging in the impossible extension of day. When I rattle up Cockpit Steps, I slink from light to dark to light again within a few paces. The next night, I'm struck by the flat open dark of Horse Guards Parade, like an absence, an open hole in the architectural weave. Along Whitehall I step from one beam of white streetlight to the next; in the recesses of the building, tucked in the arches and behind the pillars, the black appears absolute.

Such contrasts are crucial to the way in which we experience the dark. In the visual systems of all the vertebrates so far studied – from amphibians to birds – contrast has been shown to be key to the way sight works: sensitivity to highlight and shadow are common to all of the optic mechanisms, with light and dark processed not only by separate sets of ganglion cells (located near the inner surface of the retina and transmitting visual information) but by entirely separate systems of cells throughout the retina and central nervous system, making responses to light and dark complex and subtle. We are programmed to be alert to contrast. Whenever we walk into any new environment, our field of vision changes. We're suddenly confronted with new and complex sights to take in, so our eyes and brain naturally perform a quick scan. Central to this is the search for elements of contrast, things that stand out and so might signal a threat. In a basic way, this is how camouflage works; by muting the difference between light and dark, so tricking us into thinking there's nothing to see.

In his analysis of rainbows, Aristotle proposed that dark was not merely the absence of light, but a complementary force that existed in its own right, light and dark acting like salt and pepper in our understanding of the visual world. He suggested that it was when dark and light mixed together that colour was produced – different doses of one or the other creating the spectrum of the rainbow – but that each maintained its integrity as a separate entity. As we've seen, this theory was later disproved, largely by Newton's experiments with prisms, and we now understand that light and dark don't exist as 'mixer' ingredients. But Aristotle's concept of an active tangible juxtaposition of light and dark rather than merely a state of light and not-light, seems to match much of our experience. On my London night walks, it doesn't feel as though I'm passing from light to an absence of light, from something to nothing. The darkness feels as though it has at least as much presence as the light; the dark patches and corners appear to have body, texture and substance. Dark feels real and manifest, a fitting rival to the light. Scientifically, Aristotle might have been proved wrong, but in the way our eyes and brains process the dark, his commitment to darkness as a thing in its own right appears to make sense.

In the city I move between light and dark. The historic heart of central London makes such a strong sensory impression at night because its streets and buildings set up contrast. 'How much more mysterious and inviting is the street of an old town with its alternating realms of darkness and light,' writes the Finnish architect, Juhani Pallasmaa, noticing how 'the imagination and daydreaming are stimulated' in this interaction of light and shadow. Much architecture is designed to celebrate light. Windows and skylights, terraces and balconies have for a long time been part of our quest for brighter spaces in which to live and work; 'the history of architecture,' said the twentieth-century architect le Corbusier, 'is the history of the struggle for light.' But architecture can also be designed to retain darkness, to create changes in mood, quiet spaces or seclusion. The medieval cloister, for example, provided an area for study, exercise and meditation, a place of coolness and shelter with a different feel from the rest of the monastery, where the possibility of darkness created a play of contrasts, a kind of sanctity of shadows. The movement of light and dark marked the passing of hours, the cycle of the year, the passage of the sun and moon; the architectural regularity, the patterns of repetition, allowed for the integral presence of darkness alongside light, of contemplation alongside action. Consider, too, traditional Islamic architecture, which uses the contrast between light and dark in multiple ways to create evocative spaces. The bright sunlight of hot streets gives way to the darkness of deep archways, passages and shaded patios; mosques contrive a careful balance between light and dark; the Ablaq technique of alternating rows of light and dark stone uses contrast to stunning effect in buildings such as the Al-Azm Palace in Damascus or the Cordoba Mosque in southern Spain.

More recently, the Estonian-American architect Louis Kahn, responsible for the design of Jatiya Sangshad Bhaban, the National Assembly Building of Bangladesh (1962), and the Yale Center for British Art (1969), spoke about the way darkness has the power to give form to architecture. He discussed the effects of classical Greek buildings in terms of their ability to manipulate the rhythms of contrast: 'Greek architecture taught me that the column is where the light is not, and the space between is where the light is. It is a matter of no-light, light, no-light, light ... that is the marvel of the artist.' Echoing Aristotle, Kahn believed that, in architectural terms, darkness existed alongside light, with a glimpse of light making darkness more absolute. 'Even a space intended to be dark,' he suggested, 'should have just enough light from some mysterious opening to tell us how dark it really is.'

Letting in light, creating mystery to underscore shadow and absence. Designers at the Helen Hamlyn Centre for Design at London's Royal College of Art, looking at high levels of artificial light in contemporary architecture, came to a similar conclusion about the way in which we use contrast to better grasp the dark: 'In darkness, where the information is reduced, details can only be grasped one at a time in a process analogous to touch ... In darkness the eye picks up small pieces of information; a glint of light on a polished surface, the shadow of an outline.' The idea that our eyes discover a dark environment tentatively, sensuously, like fingers edging into the corners of a cupboard, brings home the possibilities of darkness in the built environment; in this way, buildings make themselves known slowly as the eye makes sense of details glimpsed. This piecing together, a kind of visual fumbling, is described by the RCA designers as an inventive process of not-knowing and discovery which 'allows for a narrative to be constructed with a beginning and an end. It also allows for the possibility of alternative endings and other interpretations, allowing the brain to be creative in completing the picture'.

An exploration, a storytelling, a feeling your way in a 'process analogous to touch'. This kind of creative brain ties in with the process by which we often read texts or view art – we fill in the breaks and step into the incompleteness, inferring meaning from the fractured imagery of a poem, for example, or construing form in the loose, suggestive brushstrokes of an Impressionist painting. We bring our experiences and expectations, our interpretations and deductions in what art historians sometimes call 'the beholder's share'. Similarly, the incompleteness of the dark, a glint of light, allows our imaginations to ease into the gaps. Its partiality invites narrative. For the darkness to come alive to us, it requires light.

Not surprisingly, perhaps, the creativity integral to the interplay of light and dark, the potential for drama and story, has long been a source of inspiration for painters exploiting these possibilities of contrast. The Italian term *chiaroscuro,* translated as light-dark, is often used to refer to the use of bold tonal distinctions to model figures and create an impression of a three-dimensional scene that draws the viewer directly into the heart of the action. What emerges is similar to the architectural treatment espoused by Kahn and the Helen Hamlyn Centre, a vivid, lively use of contrast that relies on the resonance and wonder of the dark.

Because the invention of oil painting in the fifteenth century allowed for the blending and layering of colour, artists began to experiment with the application of light tones directly on top of dark ones to render the effect of figures emerging from shadow. Although the concept had its roots in classical Greece, with *skiagraphia* or shadow painting that used cross hatching and shading to create an impression of volume, it only really became a significant technique in the late fifteenth century, when painters began to use it to great effect to capture the passion and excitement of intense scenes, often with divine or life-and-death subject matter. Leonardo da Vinci's altarpiece panel, *The Virgin of the Rocks (c.* 1483) for example, shows the Virgin Mary surrounded by cherubic angels, their pale skin shining against a jumble of dark rocks behind. The scene, contrasting the light of divine figures with the darkness of a primordial setting, was apparently inspired by Leonardo's memory of an excursion during which he was confronted by the allure of darkness: 'I came to the mouth of a great cavern,' he recalled, 'in front of which I stood sometime astonished. Bending back and forth, I tried to see if I could discover anything inside, but the darkness within prevented that. Suddenly there arose in me two contrary emotions, fear and desire – fear of the threatening dark cave, desire to see whether there were any marvellous things within.' Light and dark. Desire and fear. Leonardo's visceral response to the 'great cavern' helps to explain why chiaroscuro continued to be a popular choice with artists attempting to provoke powerful emotions and to capture drama. The rendering of pale bodies against a dark background of shadow or foliage, or the lighting of still lives with striking shafts of light, immediately suggested intensity and magnitude: Rembrandt, Vermeer, Rubens and Velázquez were among those who adopted and developed chiaroscuro, using contrast to explore the deepest passions of the human situation, our loves, fears and tragedies, births and deaths, our relationship to God.

The name most associated with chiaroscuro is probably Caravaggio, who was a celebrity in seventeenth-century Rome as much for his tempestuous lifestyle as for his painting. Working on canvases with minimal preparation, and depicting religious scenes with the help of rough local models and contemporary backdrops, Caravaggio used a harsh light to isolate his figures and highlight emotional tension. By the end of his short career, his use of the technique had become so pronounced, with such vivid contrast of dark and light, that it was given its own name, 'tenebrism', from the Latin word *tenebrae*, for darkness. The searching light turned on the paintings' figures gave them a bold, fleshy reality but it was the strong presence of very deep, almost pure black, shadows that brought a metaphorical as well as literal darkness to the work. Caravaggio's painting of *David with the head of Goliath* (*c.* 1610), for example, shows a thoughtful, rather melancholic, young David holding up the dripping head of the vanquished

Goliath, whose final expression of anguish and despair is frozen on his pale face. The hair of both characters melds into the dark behind them, and the arm with which David holds his sword is swallowed by shadow; the characters are captured emerging briefly from darkness, as though they're about to be consumed by their deeds. This is a sombre, contemplative work in which it's the dark, rather than the contrasting highlights, that is the overwhelming presence.

Since the heyday of chiaroscuro, artists have, of course, continued to use contrast to great effect, often loaded with significance beyond its visual impact. John Constable's eighteenth-century pictures of large skies with their striking contrasts suggest a changing uncertain world. Sir Joshua Reynolds used black backgrounds in some of his portraits to hint at the darker elements of character; in his portrait of Elizabeth Siddons as *Tragic Muse*, he goes a step further, half-concealing two disquieting spectral figures of Pity and Terror in the inky shadows behind the actress's pale face and white gown. Van Gogh juxtaposes black bare trees, stumps and scribbled foliage against the whites of water and sky in works such as *The Parsonage Garden at Nuenen in Winter* (1884) or *Landscape with Bog Trunks* (1883), to draw our attention to the truths of life and death. In the twentieth century, artists explored the effects of contrast to challenge the ways we create meaning: the photographer Horacio Coppola, for example, photographed a piece of white string and a white egg on a black surface in his 1932 work of contrast, *Egg and Twine*; Bridget Riley's mid-twentieth-century compositions of black and white lines rely on tonal contrast to capture a sense of movement and, she claimed, to 'dismember, to dissect, the visual experience'. The photojournalist Arthur Fellig, known as Weegee, recorded more than 5,000 murders in New York between 1935 and 1945, using what he called the 'Rembrandt light' from the glare of his flash to set his protagonists against an otherwise black background, challenging popular views of mobsters, labour unions and downtown neighbourhoods in a series of photographs that have become synonymous with twentieth-century noir.

In the piecing together of information, the insinuation of possibility made possible by contrast, we seem to glimpse something of the brevity of our lives, the smallness of our dramas, the reverberation of moments. In a flash of light, darkness becomes visible. Chiaroscuro, light-in-dark, continues to draw us.

My dad won't turn the lights off. As his dementia grips, he sleeps less and less, wandering the house through the night in a blaze of light. When I was living at home, he always had an eye on the electricity bill; lights were not permitted unless we were present in a room. Now, he's anxious of the transition from light to dark, so he keeps all the lights on all the time, wall lights, table lamps, the television. Tonal contrast confuses and annoys him: if he comes into the garden on a fine day, he hovers on the boundary between sunshine and shadow. His brain is unable to make the creative leap required to fill the gaps left hanging between light and dark.

As other awareness slips from him, dad's recognition of contrast becomes even more acute, showing how deeply embedded in both mind and body is the opposition of dark and light. We're programmed to look for it, and many of the things we take for granted are dependent on the distinction. Why don't we usually spot stars shining during the day? They sparkle just as brightly as they do at night, of course, but without being set against a dark sky, they remain invisible to us.

Fireworks at midnight, bonfires, torchlit pageants, flares and lanterns all exploit our natural attraction to contrast. We've long been beguiled by glitter, by the twinkle of gold in the shadows, by a lit window in the vastness of the night but, these days, it can be difficult to imagine the impact of such displays on people in the past who were so accustomed to the indestructibility of darkness. Before the widespread invention of artificial light, nearly every public celebration included a night-time element, when light could be spun through the dark and people could gather to wonder at the spectacle; Barcelona was lit by fires, lamps and candles for three nights to celebrate the end of the plague in 1654, for example. In particular, such displays were used to assert power and privilege. In the Middle Ages, it was only the most affluent households that could afford extravagant lighting for feasts and revels. Holding back the dark by lighting a great hall for a night of dance, music and drink was a mark of authority. Similarly, royal courts indulged in night-time diversions because they had the wealth to banish darkness, emphasising their elite status in their taste for nocturnal entertainment. The celebration of royal births, deaths and marriages, or a significant military victory, provided the perfect excuse for the kind of brilliant spectacle which would cause the masses to gape and wonder as the night retreated. Samuel Pepys recorded the logjam on London streets on 29 May 1666, the night of Charles II's birthday, caused by the number of bonfires lit in the king's honour (a few months before the Great Fire of London); Handel's famous piece, *Music for the Royal Fireworks* was written in 1749 by order of George II to accompany the display commemorating the end of the Austrian War of Succession and the signing of the Treaty of Aix-la-Chapelle, the previous year.

Religions, too, flaunted their ability to push the dark away, at least briefly, so asserting the authority of the divine – and his ecclesiastics – over the visible world. The ancient Egyptians used light at the funerary rights of the most celebrated members of the community to emphasise their specialness: the *Book of the Dead* papyrus, from around 1450 BC, shows wicks on sticks, one of which was later uncovered in the tomb of King Tutankhamun. Candles, tapers and burners were used commonly in Catholic Europe and before the Reformation in Britain. Statues and shrines were often lit along roads or at junctions, helping people to find their way and apparently in an attempt to deter too much blasphemy or poor behaviour from passers-by. But on particularly important religious occasions, battle would commence in earnest against the dominance of the dark, often to spectacular effect. Streets were strewn with light from one end to the other, creating a radiance that would have been entirely new, and enormously impressive. Processions in Spain during Holy Week featured candles with four wicks – so burning especially brightly – and a festival in the Sicilian town of Messina in the seventeenth century was famous for making the entire town visible for miles, even in the dead of night. Processions by torchlight were frequently held on pilgrimage routes as a way of making the experience momentous, and still are. The local newspaper recorded a pilgrimage to the Catholic shrine in Walsingham, Norfolk in 1928, noting that 'at the conclusion … a procession was formed, and passing out into the dark night, singing the Pilgrims' Hymn, the Church and churchyard was encircled by a ring of light as each pilgrim proceeded bearing his candle'. Pilgrims visiting the shrine in Lourdes are invited to take part in a nightly torchlit procession at which a statue of the Virgin Mary is carried above a sea of light.

Alongside the use of actual lights to pierce the dark, many of the world's religions also draw on light as a figurative contrast to the threat of darkness. Metaphorical light in all its forms – clear, coloured, radiant, blinding – has played a central role in histories of Judaism, Christianity, Islam, Manichaeism, Zoroastrianism and Neoplatonic mysticism, as well as in Buddhist and Hindu esoteric traditions. This light is commonly associated with spirituality, knowledge, wisdom, insight, justice, and the good. But for every moment in the light, there's also a period in the dark; just as we understand the religious meanings of light, so we know to connect darkness with ignorance, secrecy, deception, corruption and evil.

But where have these associations come from? Why is it that dark and light embody such antagonisms?

If we go back as far as ancient Egypt, we can see the sun god, Amun-ra, the most powerful of gods – earlier known as Ra – waging a daily battle against darkness in the form of the serpent monster Apep, who threatens to swallow him at nightfall. Early Greek philosophers such as Parmenides and Pythagoras fostered these associations of the transcendental pitched against the bestial, writing about the ways light connected with the spirit and dark with the body. They proposed an influential theory that humans had first existed somewhere low and dark, in thrall to our brute natures and physical limitations, hauling their way to a higher level of divine spirituality which was notable for its illumination. Already we see a gulf opening between light and dark; already darkness is being lamented as an impediment to be escaped, disposed of, thwarted. Many religions picked up these ideas, repeating and adapting them. In ancient Mayan culture, for example, the Hero Twins, Hunahpu and Xbalanque, demonstrated their power through a victory over darkness and death, although periods of dark chaos were considered a constant threat to be mitigated by sacrifice. Māori oral traditions similarly describe a process of becoming that involves emerging from darkness into light. Here the dark can take many forms, from the depths of night to the darkness of things growing in the soil or the womb. As we've already seen, darkness is given form through the existence of Whiro, who is cast in opposition to Tāne, humanity, in an enduring contest between dark and light. After a gruelling struggle, it's Tāne who eventually comes out on top as darkness is banished: 'Whiro was driven down to the underworld, hence he is called Whiro ki te Po, and thus Light was triumphant in this world. Tane conquered Darkness.'

The battle between dark and light was fundamental to the influential beliefs of Manichaeism, a religion founded in the third century AD by the Parthian prophet Mani. Manichaeism pivoted on the struggle between the spiritual and the material, represented as a good world of light and an evil world of darkness. It proposed three 'moments' or periods of reality in which light and darkness vied for supremacy, concluding with the triumph of the sun, moon and stars – the embodiments of light – and the final separation of light and darkness resulting in the destruction of the world. Between the third and seventh centuries, the religion spread quickly and widely through Aramaic-speaking regions until it became one of the most common belief systems in the world, rivalling Christianity in its missionary zeal and geographical reach. The central belief in the duality of light and dark, therefore, became widespread and powerful.

Over and over again, across time and place, light and dark are flung at each other through variations on the doctrine that the universe pre-existed in an unholy state of darkness, a huge swirling force of disorder and non-being that required fixing. Creation stories brought light to this emptiness, representing life and knowledge, humankind and God. The Jewish Torah explains how the world emerged from 'the unformed and void, with darkness over the surface of the deep' when God separated light from dark. Similarly, on the first day of the Christian creation recounted in the Book of Genesis, it's the separation of light from dark that begins the process of coming into being; light becomes a symbol and embodiment of God, while dark becomes a state of being excluded from God and the life he creates. Later in the Bible, the arrival of Jesus is portrayed as part of this ancient schism, the advent of light into a dark world: the Gospel of John, in particular, sets up a series of stark contrasts to drive home the message. Jesus is 'the light that shines in darkness'. Subsequently, the good–evil metaphor is made even more explicit: 'The light has come into the world,' John says, 'and people loved darkness rather than light because their deeds were evil. For all who do evil hate the light and do not come to the light, so that their deeds may not be exposed.' Similarly, Matthew's Gospel draws attention to the eye, urging us to look on the world in the spirit of purity and light: 'The eye is the lamp of the body. So, if your eye is healthy, your whole body will be full of light, but if your eye is bad, your whole body will be full of darkness.'

Such images of light and dark became fundamental to Christian teaching and thinking over the centuries that followed. Pick up any Christian text of the last thousand years or so and you'll see the motif repeated. The Gospels tell how the crucifixion of Christ was followed by 'darkness over all the land'. The Anglo-Saxon *Dream of the Rood,* an epic poem, describes how the cross was hung on Calvary in darkness. Huguenots being persecuted for their protestant belief in seventeenth-century France expressed their despair in familiar terms in a fervent prayer: 'Our congregations have been scattered; our pastors forbidden and banished from our midst!

Our sun set while it was still day! The darkness surprised us in broad daylight!'; John Henry Newman wrote the popular hymn, 'Lead Kindly Light' in 1833, calling on the Divine light to guide us through 'th'encircling gloom' of the dark night.

In early Islam, light again became a way of expressing the transcendent presence of God. In the *hadiths*, recording the sayings and traditions of the Prophet Muhammad, paradise was often described as beaming with a bright light that cut through the darkness and might be 'caught' by Muslims on earth; some Muslims believe that in the depths of the night, God descends to the lowest of the seven skies to listen in to the prayers of the faithful. Some Sufi mystics espoused a view of the soul as a light that was attached to the dark substance of the body. Today, these long-standing Sufi ideas about God as light (*yeleen*) can be seen among communities in the Ivory Coast, for example, where scholars perform a devotional practice of concentration called *dhikr* (or *zikr*) in the middle of the night. While maintaining that God is beyond darkness and light, and cannot be confined by either, the *dhikr* brings the mystic physically into the deep dark and spiritually 'within light', the contrast between the two states allowing for spiritual enlightenment.

Like the *dhikr*, many of the rituals of the major religions pit dark against light in a physical as well as a mystical or moral sense. The Diwali festival uses lights to signal the victory of good over evil; Christian baptism surrounds the child with lighted tapers and candles to drive away the darkness of evil; most religions preserve a source of light in front of an altar. In several rites, a devil figure is driven out of the dark, considered to be the realm of his terrible activity; for feasts and special worship celebrants commonly wear light-coloured robes. Since in many religions, God cannot – or should not – be directly represented, light offers a way of suggesting the advent of the Divine into our murky existence of confusion and sin. Sunlight streaming into a shuttered room; mirrors radiating the purity of good souls; a lantern held aloft in the winter dark; all act as visual metaphors for the presence of God. In religions where more representational iconography is permitted, the light-dark opposition is so routine that we often hardly notice it: bright haloes contrasting with dark backdrops; angels sparkling against night skies; candles and fires blessed and lit in the gloom of caves or churches, the shining gold of heaven set against the dark abyss of hell.

Dante's fourteenth-century narrative poem, *The Divine Comedy*, begins in a dark forest, 'savage, rough and stern'. As the poet-narrator proceeds on his journey into the depths of hell, darkness remains a key feature: the inferno is not so much a place of fire but of the terrifying, unremitting dark. The marvellously gruesome paintings of the Last Judgement popular during the Middle Ages similarly drew on all the church's iconography of darkness; in their powerful visions of the contrast in fate between the saved and the damned, they show some figures welcomed into the light of Heaven as their reward for good lives with others punished by being thrust into the dark void of Hell for eternity. Giotto's early version, a huge fresco painted around 1306 in the Scrovegni Chapel in Padua, shows the ordered ranks of angels, apostles and saints basking in the glow of golden haloes with the illuminated figure of Christ at their centre, rays of light or fire bursting from his throne. Meanwhile, in the lowest corner of the wall, a black cavern is home to the monstrous horned devil and the tortured naked figures of the damned. Both the devil and his smaller worker demons are an ashy grey; the devil is shown energetically devouring the sinners and excreting them further into the depths of darkness.

$\star^+_+\star\ (\star^+_+\star$

Is it all or nothing then? Light or dark? Light obliterating dark? Historically, this was the view of the more rigorous proponents of Western religions acting on the basis of clear oppositions, and of many political leaders who found images pitching the power of light against the corrupt forces of darkness too tempting to resist. The Roman emperors Vespasian, Hadrian and Caligula were among many who characterised themselves as Helios, the sun god, especially on coins, all-powerful and victorious. Later, European monarchs adopted the same metaphors as they aimed to strengthen their political position and reinforce the concept of absolute power. Henry IV of France portrayed himself as 'the light of the world'; his successor liked to be known as 'the Bourbon star'. Perhaps most famously and effectively, Louis XIV, the Sun King, used the symbolism of light systematically and consistently to present himself triumphing over darkness.

But this annihilation of the dark in the dazzling worship of light is not the end of the story, a battle won, as though some kind of status quo has been safely regained. While it can be easy to dismiss darkness as nothing more than an obstacle or deficiency, the embodiment of evil, some religions, especially those from the East, regard it as part of a cycle in which dark and light, blessing and sin, good and evil are contingent on each other, complementary rather than simply oppositional. This is a view that not only allows the dark but actually requires it: just as we need a dark night to be able to see a light source, say a candle flame set at a distant window to guide us, so the light of spirituality and wisdom demands a knowledge of – and engagement with – darker states in order to shine.

Many religious teachers point out that God did not abolish darkness at the creation. Darkness remained, primordial and essential, but God added light, which, they claim, can only be thoroughly understood in contrast to the pre-existing dark. While darkness is mitigated in the Torah's creation story, for example, it's not eradicated; it remains as a counterbalance, a condition of being deprived of spiritual blessing: the penultimate plague to afflict the Egyptians is *hoshekh,* an extreme darkness, so thick and absolute it 'can be touched', wounding both physical and mental well-being. Similarly, many religious scripts return us to the dark, often portraying the spiritual journey as a passage through darkness. The glimpse of light at the end of the process is what drives the believer ever further, but the darkness that envelops them is fundamental to the process of discovery and change. One of the clearest expressions of this balance is probably in the Chinese philosophy of Tao, and particularly in the concept of yin and yang. Developed over thousands of years, and related to astronomical observation, yin-yang is a complex relational concept that perceives dark and light as interconnected and interdependent: most of us recognise the symbol, which fits identical but inverted black and white shapes into a complete circle to demonstrate wholeness. The white shapes represent the male element in the universe, associated with the light and daytime. The black yin is the feminine element representing night and darkness. Each needs the other to hold the balance of the circle. In Chinese creation mythology, yin and yang emerge from the life force, *qi,* in a transformational moment which sets up a perfect equilibrium of binary elements.

The Swiss psychiatrist and psychoanalyst, Carl Jung, working in the first half of the twentieth century, studied Eastern religious practice, including Tao philosophy, as part of his inquiry into how human psychology operates. His research into the nature of the unconscious considered the balance between what he described as the light world of our conscious mind and the dark elements of the unconscious, our more selfish, destructive and amoral impulses, which we usually conceal. Echoing the teachings of many religions, he concluded that aiming for the elimination of the dark side was pointless and reductive. More important was to keep the light and the dark poised in a kind of continuous conversation: 'I must have a dark side,' he suggested, 'if I am to be whole.' Other writers and thinkers have also emphasised this notion of wholeness, the equilibrium between light and dark, and the interdependence of one on the other. The American novelist Ursula Le Guin, for example, who wrote science fiction and fantasy work influenced by both Jung and Taoism, wrote:

> Light is the left hand of darkness,
> and darkness the right hand of light.
> Two are one, life and death, lying
> together like lovers ...

In her popular and successful 1969 novel, *The Left Hand of Darkness*, Le Guin offers an examination of androgyny, establishing a fictional future universe which challenges divisive models of gender, ethnicity and society and which is pervaded with imagery of light and dark. 'Duality is essential, isn't it?' asks one of Le Guin's characters.

The nineteenth-century philosopher Georg Hegel summed up this unity in contrast nicely in his attempt to define what he called a 'doctrine of being', published in 1812 as *Science of Logic*. 'Pure light and pure darkness are two voids,' he claimed, 'which are the same thing.' In such pure states, he said, both light and dark are 'nothing'; they require some kind of mingling or integration in order that 'something can be distinguished'. He concluded that the two hold and counterbalance each other: 'light is determined by darkness ... and darkness is determined by light.' I consider this fundamental unity of light and dark in my city night walks. I'm in New York. Specifically, I'm at the top of the Empire State Building. It was a last-minute decision to come here, after leaving an event close by; I discovered by chance that visits were permitted until late and arrived just in time to catch the last elevator to the top, dashing in from the street and straight up. There was no queue, no noise; it was rather like being late for a routine appointment. And now, standing on the viewing platform, I have the place more or less to myself; there are a couple of people the other side of the tower, but that's all. It's quite a close, cloudy autumn night but I can see the city lit below and around me; other skyscrapers brightly illuminated, office and apartment lights, the glowing grid of streets and the horizon a dense greyish sheet hung behind. I reflect on the ways in which cities like this reconstruct ancient creation myths, taking the unlandmarked dark and populating it with light; ordering the confusion with bright building blocks. I marvel at the spectacle of confident illumination. But what strikes me most is not sight but sound. A shower of twittering, like white noise. A sort of light-fog hangs around the top of the tower, a misty glow. In and out of this flutter little birds, a swirling frantic flock, a scuffle of wings and urgent chirping.

In September 1970, the *New York Times* reported that 'Hundreds of migrating warblers, thrushes, vireos and tanagers, flying low under cloud cover and confused by bright lights, crashed into the Empire State Building early yesterday morning and fell onto the setbacks or down to the street.' One of the policemen called to the scene in the middle of the night is quoted as saying 'it was raining birds'. Despite rescue efforts, around five hundred birds were found dead; many more were probably never discovered.

Most songbirds migrate at night in spring and autumn, when the air is cooler and calmer, and fewer predators are out and about. When it should be dark. But the city lights interrupt these natural events. The first recorded deaths of nocturnal migrants in New York were documented in collisions with the Statue of Liberty in May 1887, a couple of weeks after the installation of electric light. Since then, 90,000 birds are estimated to die each year smashing into New York's skyscrapers; by some estimates, over 300 million birds – or perhaps as many as a billion – die annually worldwide because the lights from tall buildings interfere with migration. In an effort to reduce the carnage, there's now a lights-out policy in some cities, including much of New York, which means the Empire State Building, and other tall buildings, try to dim or extinguish lights by midnight.

1,250 feet up on the observation deck, in the middle of the night, I am surrounded by birds trying to find their way to distant places for the winter, wheeling in the smudgy light at the top of the tower, disorientated, burning up reserves of energy. The balance of light and dark is upended, contrast eradicated. The birds' natural rhythms are out of joint, their navigation systems jammed. I walk round the gantry in a halo of illumination and I'm taken aback by the anxious noise: the flap and chitter is unexpected and memorable, otherworldly, disturbing. It reminds me how lost we are without the dark.

⋆⁺₊⋆ ☾⋆⁺₊⋆

The word 'enlightenment' parries the dark. To enlighten is to remove darkness. But it's an unusual word in English because it's always used in a figurative, rather than a literal, sense – we never enlighten a room, whereas we might enlighten an ignorant cousin.

But where does it take us, this metaphor of light superior, of the defeat of darkness?

By the end of the fourteenth century, 'to enlighten' had already cropped up in the context of relieving a heavy heart but in time, attention shifted from body to mind: by the 1660s the term was applied specifically to intellectual enlightenment, the gaining and giving of knowledge. This was the meaning that stuck. By the middle of the nineteenth century, 'enlighten' was firmly established as a way to describe the practice of shedding light on a subject from a position of understanding and through a process of rigorous inquiry. We now use it not only as a verb, of course, but also as a proper noun: the Enlightenment describes an entire European movement of the seventeenth and eighteenth centuries, in which our understanding of God and the natural world, human nature, politics, science, philosophy, art, literature and even gardens was transformed, old traditions and habits set aside in the pursuit of 'modernity'. In this context, there was no room for the equivocation of darkness: 'enlightenment' captured the mind's capacity for clear and orderly thought, true perception, and self-awareness; ideas of transparency provided the metaphorical ground for identifying things that should be widely known and shared as humankind moved towards greater personal freedom, changed political systems and democratic approaches to power. To cast light was incontrovertibly a good thing. Flawless light became an object of fascination. The dark was banished.

It's easy to see how the word 'enlightenment' came to be connected with concepts of knowledge. Firstly, there was the entrenched vocabulary of Western Christian religion which, as we've seen, associated darkness with a lack – lack of morals, of grace, of belief. The acquiring of (Godly) knowledge overcame this state of darkness; it was the light of wisdom which would pierce the gloom of sin and ignorance. At the same time, new discoveries and approaches during the seventeenth and eighteenth centuries began to unearth connections between the brain, body and light, helping us understand that reasoning and cognition – which were so highly regarded during this period – depended at least partly on the stimulus of light, as did memory and the functioning of human organs. The English Enlightenment philosopher, Thomas Hobbes, for example, wrote several works examining the nature of light, and its impact on body and mind; John Locke's inquiries into what knowledge is and how we acquire it, drew attention to the power of the senses in delivering information to the brain, and the role of light in manipulating this sensory experience through the ways in which we perceive colour. Isaac Newton's *Opticks* (1704) examined theories of light and optics; there was widespread enthusiasm for scientific instruments such as telescopes and microscopes that drew direct connections between light, eye and understanding; medical knowledge of the brain moved rapidly, through research by physicians such as Thomas Willis, who published his *Anatomy of the Brain and Descriptions and Functions of the Nerves* in 1664. From these experiments on the physical nature of light it was a relatively simple step to more conceptual discussions of the nature of 'enlightenment'.

But all this dwells on an enduring and increasingly informed preoccupation with light. My interest is in the dark. What of that?

The answer might lie again in the significance of contrast. Because in its quest for illumination, the Enlightenment drew on darkness, frequently plunging into the vocabulary of the dark in an attempt to express its beliefs and define its difference. Motifs of light and darkness, which had been the mainstay of religious imagery, were tweaked for a new era and to reflect increasingly secular attitudes: now it was scientists and philosophers, rather than God, who were shown to be the source of light; darkness became a state of superstition and prejudice rather than moral failing. But despite this change of emphasis, darkness remained all around. The Enlightenment needed the dark for the light to blaze. 'At last, all the shadows have been banished. What a light shines from all sides!' exclaimed the French statesman Anne Robert Turgot in 1750. Around the same time, the French writer, Voltaire, who was sharply critical of Christian, especially Catholic, practice, called on right-thinking citizens to 'pass through the darkness of ignorance and lies' so they could return 'to your palace of light'.

It was by drawing on the dark that the revolutionary thinkers and speakers defined the new way of doing things. Those who were seen to prevent inquiry, experiment or the diffusion of knowledge were termed 'obscurantists', clinging to the dark and cloaking the light. Similarly, the French word *ténèbres*, literally translated as darkness, came to describe a state of intolerance and despotism, and those who espoused it. *Eteignoir*, also from the French, and meaning to extinguish, came to be associated with churchmen hostile to the Enlightenment and political reactionaries who wanted to snuff out new libertarian movements; *les éteignoirs* were often caricatured with long clerical candlesnuffers, extinguishing the bright lights of justice and reform.

The public battle of light and dark during the Enlightenment has endured: in their 1990s book, *The Demon-Haunted World: Science as a Candle in the Dark*, for example, the American astrophysicists Carl Sagan and Ann Druyan drew on remarkably similar phrasing to address the conflict between knowledge and superstition: 'Pseudoscience and superstition will seem year by year more tempting, the siren song of unreason more sonorous and attractive... The candle flame gutters. Its little pool of light trembles. Darkness gathers.' But the language of opposition and contrast has perhaps never again been as keenly employed as it was towards the end of the eighteenth century, when it seemed to express the impossible certainty that the dark really would be forced to retreat and night might no longer follow day. That many of the terms being flung around were French was of course no coincidence. New discoveries and ideas, a changing understanding of the relationship between the individual and the state, and rigorous public discussion of politics, religion and society, created the conditions that inspired both the long drawn-out misery of the American Revolution of the 1870s and '80s, and the brief, violent turmoil of the French Revolution in 1789 – a revolution which was frequently expressed as a conflict between light and dark. 'Despite the obscurity of the twilight in which the nations still seem to be wandering, the frequent flashes of light herald the dawn and the coming of day ... the darkness dissipates,' exclaimed the French philosopher Baron d'Holbach in his *Essai sur les préjugés*, in the lead-up to the Revolution. 'A multitude of rays gathered together will some day form an immense multitude of light that will warm every heart, that will illuminate minds.' 'Flashes of light', 'a multitude of rays', illumination – it's a familiar turn of phrase: new, progressive thought getting its day in the light while old ways and ideas are dismissed as 'the darkness'.

While it might be a far stretch to blame the dark for the French Revolution, it's worth dwelling for a moment on the dogged use of contrast – the linguistic head-to-head of light and dark – as a means of fuelling antagonism. The adoption of the term 'enlightenment', and the consequent insistence on the primacy of light, helped foment the basic dichotomy which proposed light as a solution and salve, a vibrant antidote to all the evils of darkness.

It was during the Enlightenment period that the phrase 'the dark ages' came into common use, a vague term that threw together more or less anything from the millennium leading up to the Renaissance, denigrating entire civilisations of the past: the massive and influential work, *The History of the Decline and Fall of the Roman Empire* (1776-89) written by the English historian, Edward Gibbon, is just one example from a body of discourse that laid into 'the rubbish of the dark ages'. This demonisation of 'dark' states of ignorance and apparently uncivilised behaviour, was not only presumptuous and unjustified – clearly many beautiful, clever and skilful things emerged during the Middle Ages – but also dangerous. It set up a simplified them-and-us dynamic that helped to make anything that was different seem inferior and threatening. In the brusque employment of metaphor, objects, philosophies, beliefs and peoples from a different time or place, class, context or culture could now be dismissed in the sweep of the dark.

Entrenching the idea that 'darkness', in word and idea, could be used as shorthand for everything unusual, unknown, or simply not the same, as well as for all things backward and bad, was to have far-reaching consequences. The Enlightenment was also the period when exploration took off, with travellers from the major European powers forging new routes for trade, religion, natural resources, colonial ambition, or simply for curiosity. As these travellers discovered the world, it was all too easy to make metaphors of light and dark manifest: here they were, venturing from a place of intellectual brilliance into the deep dark of the unknown. All too quickly and thoroughly, the ideas of darkness they espoused came to be associated with particular places and, above all, races. Christian missionaries, for example, sent out to follow colonial governments in Africa and Asia, were quick to adopt the language of darkness in the certainty of their creed.

These missionaries believed those without faith were 'blind', unable to see the truth of God, and confined to a state of ignorance. They were living in darkness. Take this, for example, a Welsh missionary hymn of 1772:

> Let the dark benighted pagan
> Let the rude barbarian see ...
> Kingdoms wide that sit in darkness –
> Grant them Lord the glorious light.

Or this missionary hymn written by the Sheffield curate Thomas Cotterill in 1819:

> O'er the realms of pagan darkness,
> Let the eye of pity gaze;
> See the kindreds of the people
> Lost in sin's bewildering maze; -
> Darkness brooding
> On the face of all the earth.

For these Christian travellers, the world beyond the lit churches of their faith was huge, obscure and scary, populated by an undifferentiated mass of 'pagan' or 'rude barbarian' peoples sitting in darkness, waiting to be saved. Missionary texts were full of references to what William Ellis, a Christian missionary to Madagascar called, 'the light of the Gospel' which 'could dispel the darkness' of unbelievers.

Missionary texts, letters and lectures played a large part in constructing ideas about land and peoples outside Europe. Africa, in particular, was presented to a curious European public as wild, godless and unboundaried, a place that tested the brilliance of an enlightened society peering back at it. As trade with Africa developed during the sixteenth to eighteenth centuries, it remained an ambiguous idea, a place of mystery, danger and even monstrosity, but not yet necessarily caught up in all the implications of darkness. With the rapid expansion of colonialism in the late-eighteenth and early-nineteenth century, however, narratives of darkness became useful, not just to missionary endeavours but to traders, administrators and politicians. Adopting the language and mythology of a 'dark continent' allowed Africa to be neatly positioned as a place of homogeneity, an errant outpost that required taming and civilising, and so colonial frontier spaces were increasingly portrayed as sources of superstition, savagery, irrationality – and darkness. An innate darkness that was rendered visible in the colour of the skin.

As this concept of a dark counterpart to European society took hold, difference in skin colour acted conveniently as a visual shorthand which separated the ruler from the ruled. With this leap, not only was racial difference and belonging consolidated, but the association of dark skin with dark character or subdued intellect was all too easily established. In *The Races of Men* (1850), the influential Scottish anatomist, Robert Knox, summed up this attitude: 'I feel disposed to think there must be a physical and, consequently, a psychological inferiority in the dark races generally,' he wrote. According to Knox, the characteristics of the 'dark races' were natural and immutable. And many took his word for it: dark on the outside, dark on the inside.

Works such as *The Races of Men* allowed for easy assumptions and simplistic assessment, but the associations it articulated had been a long time forming. Centuries of commentators had made a case for the intransigent connection between dark within and dark without. Greek and Roman medicine asserted that noxious black bile altered both outward physical appearance and the body's internal organs, as well as causing emotional distress and psychological disorder. Centuries later, medieval communities believed that the Black Death was a divine punishment, a manifestation of the darkness of sin inherent in humankind. On a similar basis, black skin was frequently regarded as a mark of damnation, an argument used in some quarters to justify the launch of the slave trade in the mid-fifteenth century. A few centuries later, the eighteenth-century German doctor and painter, Carl Gustav Carus, produced influential works on the nature of race, which confidently proposed a hierarchy based on the assertion that those of darker skin, *die Nachtvölker* (literally, nightfolk), were *einer niederen Individualität* (a lower individuality). By the time Henry Morton Stanley published his enormously popular accounts of his expeditions in *Through the Dark Continent* in 1878 and later, *In Darkest Africa* (1890), the language of the dark had been systematically assigned to Africa and the implications of black skin thoroughly embedded in thinking and writing. In a changing world of expanding ambitions, mass industrialisation and social upheaval, the association of the 'dark races' with the 'psychological inferiority' identified by Knox consolidated arguments for intervention and redemption by white imperial powers. At a time when taxonomies of the natural world – especially the work of Carl Linnaeus in the eighteenth century – were working to classify species of animals and plants, positioning light against dark, white against black, also enabled humankind to be conveniently ordered and ranked, so suggesting that we could be understood and regulated simply by applying a structure predicated on contrast.

Theories, histories, problems and dialogues of race are huge and complex, and beyond the scope of my inquiry into darkness. But I don't want to ignore the fact that language and ideas of the dark are inextricably bound to language and ideas of race; I can't overlook what happens if we talk about a person's skin as 'dark'. As a white woman with a privileged upbringing, my experience is narrow and my encounters with these connotations of darkness impersonal and mediated; I don't know what it feels like to be described as 'dark' with all the tussles of meaning such a descriptor exposes. But this is a chapter exploring the notion of contrast, the setting of light against dark, and it would be thoughtless not to consider, as far as I can, what this might mean in relation to race and skin colour.

I'm going to begin with wide-brimmed hats. With parasols and awnings and an obsession with pallor. Because, of course, at the same time as meanings about dark skin were cementing, there was an evident passion for keeping pale skin as pale as possible. White women of the middle and upper classes, in particular, threw themselves into battle against the effects of the sun. Natural shade or confinement indoors was rarely enough: in their efforts to avoid any taint of darkness, they used a range of beauty products to enhance their whiteness and conceal blemishes such as freckles. Well into the twentieth century, newspapers, magazines and hoardings in the West regularly advertised powders that would conveniently hide hints of a tan, as well as numerous bleach treatments. Bleachine Cream, for example, by the cosmetics brand Elizabeth Arden, was given prominence in an issue of *Vogue* in 1920, where it was lauded as 'a mild but effective preparation for removing tan. Nourishing as well as whitening.'

The desirability of pale skin travelled with other imperialist ideals, so that as Europeans colonised new countries and constructed new social norms, they took with them the cosmetic

practices of whitening and lightening. Skin-lightening prepara-
tions had been popular outside Europe for centuries, especially
in China, Japan, Persia and pre-Mughal India, a way of marking
status and defining beauty; the Romans used white lead mixed
with rose oil, and in Britain, Elizabeth I famously presided over a
court of painted white faces. But with the colonial impetus, the
attractiveness of pale skin reached new peoples and accrued new
significance. Rank, privilege, marriageability, refinement, education,
differentiation, a longing for home could all be read in the pallor
of a woman's face; cookbooks and medical manuals were packed
with hints and recipes for obtaining the most effective treatments
especially for those sequestered in remote stations of Empire and
having to make do with home-brew concoctions (similar recipes
still occupy large amounts of 'how-to' space online). Arnold Cooley,
a nineteenth-century British surgeon, for example, explained that
while some women were content with using starch or finely
powdered carbonate of magnesia, for those in search of 'greater
whiteness and brilliancy ... pearl-powder (subchloride of bismuth),
pearl-white (trisnitrate of bismuth), and hydrated oxide of bismuth,
hold the first place. Sometimes even flake-white (carbonate of lead)
and white precipitate (ammonio-chloride of mercury)'. He noted
that plastering the face in such poisonous substances often led to
a 'leathery appearance' or 'partial or complete paralysis' but such
consequences were rarely enough to disconcert users, and products
containing mercury, arsenic and lead were commonplace.

Such unpleasant and hazardous treatments were not confined
to white daughters of Empire. Women of other races were (and
in places still are) keenly alert to gradations of darkness; tones of
dark skin could be rated and judged; paleness stood in contrast not
to a simple rendering of 'dark' but to a minutely ordered strata
predicated on apparent agreement about how much darkness was

tolerable. Another advert for Elizabeth Arden's Bleachine Cream appeared in the *South China Morning Post* in 1937, for example, while Indian society has long been painfully alert to the nuances of skin colour: the World Health Organisation estimates that the industry for skin-lightening products in India is currently worth over $500 million. In the USA and South Africa, too – where it could be argued that the contrast between dark and light was most structurally embedded – black women were routinely the subject of marketing campaigns for skin whiteners. Sometimes these women had been mission-educated, and so were profoundly aware of the religious associations inherent in ideas of the dark; sometimes they just wanted to be fashionable and fit in. Whatever the motivation, the implications were significant, driving home the grave importance of visual indicators of race – above all, the darkness of the skin – while insidiously privileging whiteness. In the comparison of impeccable white skin with undesirable darkness was the proposition that dark skin was somehow a choice, something you could acquire (through tanning) or disown (through whitening), and that in opting for a particular skin tone you could also select race and belonging.

So what does this preoccupation with skin colour tell us about the dark? Or is the better question, what manner of darkness do we think we can see on the skin of the people around us?

In her essay, *Playing in the Dark*, in which she traces the omissions, contradictions and conflicts through which 'a real or fabricated Africanist presence' became crucial to American writers, the Nobel Laureate, Toni Morrison, draws our eye back to the notion of contrast. It's in the polarity of skin colour, the setting of dark against light, she suggests, that concepts of the 'other' fully emerge; looking at people of different skin allows for 'the projection of the not-me'. She draws attention not just to black skin, which she terms the racial object, but also to white, the racial subject, the oscillations between them and the establishment of difference. Morrison returns to the language of the dark to express the ways in which we learn about, and define, ourselves through comparison. This process, she claims, is 'a means of meditation – both safe and risky – on one's own humanity'. Reminding us that darkness is conveniently blank, she describes how this has historically allowed white anxieties, preoccupations and ideas to be transferred onto it. In America, she claims, this has resulted in 'a fabricated brew of darkness … a reined-in, bound, suppressed and repressed darkness'. This swirling impressionistic dark of appearance and meaning, intimacy and distance, the obvious and the unconscious, coalesced into a rendition of black peoples as portrayed by white writers from Poe to Hemingway, manipulated over time into a persona sitting in contrast to the construction of the white man's self. This dark presence Morrison describes as 'a disrupting darkness', a challenge and an unease made evident in the colour of skin.

Morrison's 'fabricated brew of darkness' is a heady concoction. She traces how it enables the separation and consolidation of identity, a way of approaching abstract notions like freedom, morality, aggression, sexuality, society, power and hierarchy by trying them out at a distance. At times, as she notes, the dynamics of the contrast between dark and light grind towards suppression and repression, disquiet and anxiety, but the interaction can also be more lithe and beneficial. In the imaginative encounter with other races – in literature and art, in film and music – we're able to (indeed, drawn to) think about ourselves at the same time as those who are different to us. Dwelling on distinction, we're able better to establish who we are: 'to display knowledge of the other,' Morrison notes, is 'to ease and to order external and internal chaos' in our own lives.

In our summoning of darkness, the projection of white interiorities onto black skin, Morrison identifies all manner of fears. In transposing some of their deepest terrors onto other (especially enslaved) populations, Europeans were able to stand back and explore the things that most alarmed them, facing up, at one remove, to 'their dread of failure, powerlessness, Nature without limits, natal loneliness, internal aggression, evil, sin, greed'. While imperial forces may have shifted, we can see how this same tendency to project into the blank of darkness remains. Contemporary British newspapers continue to happily and routinely bemoan the destabilising and menacing 'dark forces' unleashed by refugees and migrants, for example, and our reading of 'dark' skin is consistently shown to prompt instinctive reactions influenced by our conscious or unconscious fears. Investigation of photos used to illustrate news stories featuring celebrities or politicians in the USA between 2011 and 2013 found that articles containing negative content were more likely to be accompanied by darker coloured images. When the press wanted to splash a story of misery and misconduct, anything from corruption to marriage break-up, they drove home the message by artificially darkening the skin tone of the protagonists. This was shown to be a constant, whether the subject was a victim or a perpetrator, and whatever their gender or race.

In another experiment, participants were given grainy video surveillance images of two men and provided with a brief sentence describing what each was doing either before or after the picture was taken. For one image, they were told that the man had done something virtuous, risking his life to save someone, or establishing a charity for children. For the other, they were told the man had committed a serious crime, such as murder or abuse. After viewing the video and descriptions, the participants were shown headshots of two completely different men. One had been artificially darkened and the other artificially lightened. When asked to identify which was the bad guy in the video footage, most chose the darker face. In a disturbing twist, participants were then invited to indicate the 'colour of the soul' of both men. Using a spectrum ranging from black to white, participants clicked on the shade that seemed to them to represent the appropriate darkness at the core of the men's being. Racial prejudice was taken into account and controlled for. Despite this, most participants rated the man with darker coloured skin as having a darker soul – even without direct reference to race, the assumption was still that the blacker the skin, the more intense the propensity to evil.

Such responses are completely automatic. We make them without being aware of what our brain is doing as it works to process differences in dark and light and make complex predictions about what this might mean; we're generally blind to the pitfalls of perception. Nonetheless, we draw on information buried deep in our neural archaeology to interpret in a particular way the sensory data provided by, for example, some video surveillance tapes. Unconsciously, we make assumptions about the dark.

Why is it so difficult for us to let go of what's surely an outdated connection between evil and darkness? What is it about the dark which continues to repel?

We're tangled in this mess of suggestions and insinuations, of assumptions and inuendoes. Darkness takes us to awkward, uncomfortable places. Dark spaces, dark faces, crime, racism – there are so many nuances pinging around, like rubber balls fired off in an empty room. Which brings me to the question of sex. Darkness as an expression of eroticism; our sexualised bodies in the dark. Another case of shadows and contrasts, of obscurities. Where to begin?

Shall we go back to black skin? At least for a moment? Because it seems important to acknowledge the powerful historic gaze of light on dark, of white skin on black. There's a long and painful history of white men fetishising black women, of plantation owners and enslavers sexually abusing men and women because darkness could be seen as erotic and exotic, as sexually alluring, as animalistic. As we've seen, there can be an inbuilt tendency to view darker skin as an indication of propensity to crime or the failings of the soul, of moral depravity. This slides us down the slippery slope of unpleasant assumptions to the point where black women are viewed as having lower moral standards than white, allowing them to be presented as hypersexualised, with an insatiable appetite for sex, or as objects for gratification. Women's black skin can all too easily mean they're presented as either sexual predators or as victims, their 'darkness' acting as an allure but also provoking revulsion.

So there are plenty of illustrations (should you wish to search them out) of enslaved women – and occasionally men – facing multiple forms of sexual violence and coercion; there are many moving accounts of the sale of strong young black enslaved people at auction for the purposes of 'breeding'. Nor was such behaviour confined to plantations. The eighteenth and nineteenth centuries witnessed a prurient fascination with the sexual lives of the peoples 'discovered' during colonial expansion. The diary of a Christian missionary to New Zealand in the 1820s, for example, noted that many Māori women, naked or scantily dressed, with loose hair and free movements, were simply 'too wild and dirty' to allow in the house. African women were consistently referred to in explicit sexual terms: William Bosman, a merchant for the Dutch West India Company (thoroughly embroiled in the trade of enslaved people) described women from the Guinea coast as 'fiery' and 'so much hotter than the men'. Another trader, William Smith, admired African women for being 'hot constitution'd Ladies ... continually contriving stratagems how to gain a lover'. The spectacle of tribal women was frequently offered as entertainment in eighteenth- and nineteenth-century travelling exhibitions – or freakshows – which allowed for the scrutiny of darkness from a safe distance. Sarah Baartman, who was from the Eastern Cape of South Africa, became known as 'The Hottentot Venus' when she was exhibited across Europe more or less naked around 1810, so that those who paid two shillings to see her could wonder at her 'very singular appearance': she was famous for her large buttocks and the object of much speculation regarding the exact nature of her genitalia.

But sex and the dark is not just about skin colour or race. We've seen how darkness creates mystery, unease; it stimulates a direct response in the body as well as the mind. The perfect conditions for eroticism. How many of us dim the lights for a romantic evening? How many sex scenes on film take place in a secluded spot or are shot in a shadowy half-light?

At the most basic level, the pairing of sex and the dark is purely practical. Darkness allows us to engage in intimacy without being seen. In times and places when privacy was at a premium, it was the dark corners, squares and lanes which allowed couples to come together for a spot of romance. When a writer for *Town and Country Magazine* recorded his impressions of Rome for an article in 1792, he described being struck by the darkness of the city at night, and the opportunities this provided for sexual encounters. 'In the midst of this darkness, amorous assignations in the street are not uncommon,' he wrote, explaining how passers-by were usually met with the command '*volti la luce*', meaning turn away the light, so that the clinched couples could remain obscured.

In England, 'dark cully' was a popular slang term during the eighteenth and early nineteenth centuries for a married man who only visited his mistress under cover of darkness, for fear of being found out in his extramarital exploits. On a June evening in 1668, Samuel Pepys – a man who frequently indulged in sexual affairs – explained how, despite the long hours of daylight, he waited until it was dark before heading off to 'la house de Bagwell' and 'after a little playing and besando, we did go up in the dark *a su cama* and there *fasero la grand cosa* upon the bed.'

But is there something other than the pragmatic arrangements for romance that draws us to the dark? It seems such an obvious connection that it hardly requires a second glance. Privacy is made possible in darkness; leisure is likely when the day's chores are finished. But neither privacy nor leisure are essentially erotic. So are there fundamental qualities of the dark itself which stimulate mind and body?

Let's enjoy, for a moment, the seductive figure of the Queen of Sheba. First appearing briefly in the Torah as a wealthy and enigmatic visitor testing King Solomon with questions, she's grown from this throwaway mention to become a cultural icon. She appears in the story of Bilqīs in the Qur'an, the epic Ethiopian foundation story, the *Kebra Nagast*, and in Persian tradition. But the fascination with her as a woman of mystery and sexual intrigue goes far beyond these cameo roles. With no evidence as to who she was or where she came from, she's provided an enticing blank canvas; over time, fanciful detail has been added to the fragment of her initial appearance, fleshing her out, and loading her with the desires and fantasies of the (mostly male) viewer. As part of this eroticisation process, she's accumulated the dark.

Sometimes she's depicted with black skin, but not always: the darkness that surrounds the Queen is more to do with her 'dark' spirit and potent sexuality. She's a woman of regal and sexual authority – she's become a temptress, attractive and inscrutable. She turns up everywhere. She features in music, most famously in Handel's 'The Arrival of the Queen of Sheba', which is now a popular piece at weddings as the bride arrives to delight her new husband. You can spot her in the great medieval cathedrals at Chartres, Reims, Strasbourg and Wells – among others – and in Renaissance frescoes at the Vatican. She's the subject of many paintings over many centuries, often highly sexualised and semi naked. In Edward Poynter's *The Visit of the Queen of Sheba to King Solomon* (1890), for example, the young queen is bare-breasted, dressed in a figure-hugging cloth and adorned in jewels, caught in the moment that she goes to prostrate herself before the King, a packed court watching the spectacle. The palace is a place of shadows, the bright day outside only glimpsed through the roof above; it's an evocative chamber of gloomy depths and half-hidden galleries. In this typically erotic, and eroticised, portrayal, the young queen emerges from the dark, a figure of promise and desire.

Mystery, sex and the dark. It's a titillating threesome. In Shakespeare's 'Dark Lady' sequence of sonnets we're again beguiled by an anonymous, enigmatic woman (or perhaps man), who remains unidentified despite the best efforts of scholars, a figure of 'black beauty', highly sexual, the object of 'lust in action.' Later, Gothic literature delved into the strange and secret lives of apparently respectable women, who are transformed under cover of darkness into highly sexualised and unknowable beings. These works simultaneously celebrated and recoiled from female monsters who transgressed norms of accepted sexuality as they revealed a darker side. Anne Bannerman's early nineteenth-century poem, 'The Mermaid', for example, features a woman-animal hybrid, strange and supernatural, keeping watch from the cliffs 'which darkness shrouds'. She lures sailors gleefully to their deaths with a mysterious power that comes to her when the wizard calls on her 'alone, by night, in darksome cave' with all the sexual implications of this. Similarly, nineteenth-century female vampires – Carmilla in Sheridan Le Fanu's 1872 novella of the same name, and Lucy Westenra, from Bram Stoker's 1897 *Dracula* – are transformed in darkness from upright women into erratic, uncontrollable, menacing monsters, and in the process become highly sexualised. Stoker writes how Lucy's 'sweetness was turned to adamantine, heartless cruelty and the purity to voluptuous wantonness'. The popular enthusiasm for Sensation Fiction in the 1860s, was equally an enthusiasm for reading about the mysteries of female sexuality laid (sometimes literally) bare. Sensation novels such as Wilkie Collins' *The Woman in White* (1859) conjured shadowy worlds of shuttered rooms, dark alleys and night-time graveyards where lone women wandered, mysterious and dangerous. Many commentators believed that such literature threatened to erode moral standards and upset domestic tranquillity by disclosing the reality of female desire and releasing the potential of darkness to overwhelm daytime constraints: the Scottish novelist Margaret Oliphant noted disapprovingly in a review of 1867 that 'what is held up as the story of the feminine soul as it really exists underneath its conventional coverings is a very fleshy and unlovely record'.

The dark, then, is a place where women's 'purity' is corrupted into 'voluptuous wantonness'. The female monsters of the nineteenth century were released by darkness, casting off their inhibitions, revealing their true, sexual natures as night fell. A few decades later, in his 1926 book, *The Question of Lay Analysis*, Sigmund Freud famously noted that women's sexuality was strange and undiscovered, concealed – and dark. 'The sexual life of adult women is a "dark continent" for psychology,' he commented, drawing on metaphors of African exploration and colonial rule, while suggesting that a woman's sex life was somehow out of reach of reasonable analysis, a place where you might not want (or dare) to venture. We're probably not unused, even now, to this idea that women's sexuality is something murky and inexplicable. It certainly proved a powerful motif through the middle of the twentieth century, as Freud's theories held sway. In Tennessee Williams' mid-twentieth century play, *A Streetcar Named Desire* (1947), for example, we're introduced to the nervy and manipulative Blanche DuBois, who is never seen in daylight. She's a woman of the dark, meeting with her lonely suitor, Mitch, at night, in dimly lit places. She shades the light in her house with paper lanterns, ostensibly to mask wrinkles on her face and hide her aging, but also in apparent recognition that she is a 'shamed' woman, exploiting her sexuality to find her way in the world when she's lost her home, fortune and respectable marriage. The implication is that a woman does best to hide her sexual activity, to keep her desire in the dark.

The tangle of sex with shame, eroticism with deviance, means that darkness inevitably lurks. In his ambitious, four-volume *History of Sexuality* (1976), the French philosopher and historian, Michel Foucault, claimed that during the nineteenth century, an obsession with formulating a 'truth' about sex meant that 'sex gradually became an object of great suspicion; the general and disquieting meaning that pervades our conduct and our existence, in spite of ourselves; the point of weakness where evil portent reaches through to us; the fragment of darkness that we each carry within us ... a fear that never ends.' In this argument, sex not only needed to be hidden by darkness, but it was also conceived as a thing *of* darkness: 'the fragment of darkness that we each carry within us.' Foucault's analysis of Victorian attitudes identifies both sex and darkness as fundamental to our existence, as powerful natural forces, as bedfellows in our psyches. But also as 'a point of weakness' that allows civilised constraints and conduct to be thrown off 'in spite of ourselves'. Here, sex and darkness come together, both threatening, both predators, irresistible and irrepressible.

In 1781, the Swiss artist Henry Fuseli, painted *The Nightmare*. It shows a young woman in a white diaphanous robe asleep – or unconscious – draped on a bed, her arms thrown behind her lolling head in a loose gesture of abandon. Sitting on her stomach, and turning its head to look directly at the viewer, is the dark, hunched, hairy figure of some kind of troll, with large, pointed ears and a grim, downturned mouth. Towards the back of the painting, an oddly theatrical (and equally demonic) horse's head looms through a curtain, its blank white eyes staring towards the woman's body. Fuseli uses chiaroscuro to powerful effect: this is a painting throbbing with darkness, the sleeping figure only momentarily snatched into the light.

*The Nightmare* proved a controversial hit. Exhibited at London's Royal Academy in 1782, it attracted over 12,500 more visitors to the show than in the previous year. Fuseli followed up by painting another version, and having prints made of the original. The work brought him enduring fame and prestige, prompting all kinds of copies and parodies. It's thought to have influenced Mary Shelley in her writing of *Frankenstein* and intrigued psychoanalysts of the twentieth century, such as Carl Jung. The blatant eroticism of *The Nightmare* beguiled and shocked visitors to the Royal Academy. The figure squatting in the centre of the painting – and challenging the viewer with its blunt gaze – is an incubus, a demon who takes male form in order to jump onto sleeping women and have sex with them.

The incubus is a predator, first appearing in ancient Mesopotamia, about 2,500 BC, but also featuring in various forms in early Christian tradition, German and Swedish folklore, as well as South African and South American mythology. It's a creature which stalks the night in search of nubile young victims to rape in the dark. Fuseli's daring choice of subject was knowingly sexy, an interpretation of night that drew openly on the multiple subtle insinuations we bring to darkness: seduction and repulsion, sensuality and violence, eroticism and deviance, the virgin and the demon.

The association between the dark and the erotic has proved tenacious. Even with more modern attitudes to sexuality, we're used to seeing sex and darkness sold as a pair. A seemingly inexhaustible enthusiasm for the modern Gothic means vampires have rarely been so visible, filling books and TV series with what the actor Johnny Depp has called 'this darkness, this mystery, this intrigue ... the erotic nature of the vampire.' Erotic computer games – admittedly a niche enterprise – are marketed with names like *Lust for Darkness*. Two hundred and fifty years after Fuseli's bold evocation of sex and the dark, the Irish novelist Niamh Campbell wrote in the *Guardian* newspaper that, 'In all the best erotic writing I've read ... a certain pleasant darkness is revealed.' But it's not just a case of complete black-out, of course. As Fuseli's painting shows, the 'pleasant darkness' of eroticism works by allowing a glimpse of something in the light – the prone figure on the bed, the glow of candles or fabrics, the shine of skin. The allure of the dark lies in the textures it offers to our senses in contrast to the glare of the day, its richness and suggestiveness; it's what we catch sight of in the dark that is seductive, rather than the impenetrability of the dark itself. In Japanese tradition, the geisha women of Kyoto worked in the shadows, inhabiting the innermost rooms of lightless buildings, but they powdered their skin to glow white and gleam in the dark. Even more distinctively, they used an iridescent green lipstick to make their mouths glisten, originally blackening their teeth to make the effect even more pronounced. Such a woman, Tanizaki explains, was a woman of darkness: 'darkness wrapped her round tenfold, twentyfold, it filled the collar, the sleeves of her kimono, the folds of her skirt, wherever a hollow invited.' But her eroticism lay in the effect of contrast, what he calls the 'visible darkness' of flicks and shimmers, of insinuation and possibility, sparks of light and colour momentarily inviting in the gloom.

Visible darkness. Not the stretched unknown of a dark night outdoors but the inhabited dark of internal spaces and whispered movement. This is where dad seems to live now, in this state of visible dark, the spluttering dark-light materiality of a world experienced through glimpses and snatches and fragments, fleetingly revealed, barely grasped before it vanishes; illuminated splinters of the past thrust into the present. This is not only a change in his mind but a new guttering physical reality. The solidity of things around him is no longer what it was; in the coming and going of the dark, objects lose their shape and solidity. He struggles to keep a grip on a chair so he can sit down; marooned in the glare of a sunny garden, the lawnmower mutates. For him, this new world is an uneven place of unsteady ground. The quivering dark trips him up. Quick, sudden contrast surprises and disorientates. He often stumbles, sometimes falls. I don't grasp what it might be like to live in this constant back-and-forth flicker between light and shadow, dazzle and dark, the unremitting ambiguity of visible darkness but I try to understand as best I can, and in doing so I'm taken back to a tunnel under the Peak District hills and the rattle of a train underground.

In my early working life, I undertook a daily commute by train. The route ran from Sheffield to Manchester, and was marked by a series of tunnels. One of these, the Totley tunnel, an astounding piece of engineering built at the end of the nineteenth century, is almost six kilometres long. The trains are local chuggers, so it takes quite a long time to travel through the tunnel from end to end; on several occasions, the train came to a halt somewhere inside and would tick over in the dark for as long as an hour.

A popular guide to the Victorian railway system, Edward Bradbury, described the experience of travelling by train through a tunnel in his *Midland Railway Sketches*, drawing on Dante's vision of hell to evoke the short Ampthill Tunnel near Bedford: 'The darkness,' he suggested, 'might be of the valley of the shadow of death – the clammy coldness ... the flakes of fire ... that flutter along, and reveal the wet walls flying past like a rushing river.' Bradbury's steam-train experience, with its flakes of fire, was clearly more visceral than my daily rattling journey behind a diesel engine. The furious glimpses of flame and water he describes throw the dark into relief, making it come alive like Tanizaki's 'visible dark'. But even in its more mundane form, the bemusing sensation of travelling through a long deep tunnel remains animated. Inside a modern, lit and closed carriage, the 'wet walls' that kept Bradbury at least a little grounded, become invisible. Beyond the train, there's only darkness. But it's a live, dynamic darkness, 'rushing' like Bradbury's river. It feels as though we're suspended in the light, perfectly still, while the darkness gallops alongside. It feels like both flying and sinking. When I looked into the dark on my way home from work, I could only see myself, reflected, seat backs, bag handles tipping from the luggage rack, bits and pieces of other passengers. The detritus of everyday travel, trapped in time and space. As though nothing else existed. As though we really had been swallowed by the earth, or, as another early railway commentator suggested, drifted back through the ages, 'like a flight through the realms of Chaos and old Night'. And because of the extent of the tunnel, and because I repeated the journey regularly for several years, this wasn't a momentary experience, but rather a lengthy repeated encounter with the clash of light in dark. Sitting in the bright tube of the train inside the dark tube of the tunnel within the unknowable darkness of rock, subject to both stillness and motion, I am in each tunnel moment both alive (of course) and also on the brink of death – a tiny fluctuation in conditions, a fleeting overturning of this particular arrangement of engineering, nature, speed, material, dark and light, and everything would be quite disastrously different. Equilibrium is all.

Darkness is very visible in a train tunnel because it contends with vestiges of light. It is, too, a necessity. The darkness of the tunnel holds me in time and space; it makes possible the passage of the train underground. Without it, I wouldn't get to work.

Māori culture speaks eloquently of this necessity. If the Enlightenment suggests a permanent state of light has been achieved and should be maintained, Māori oral traditions suggest otherwise. They draw attention, instead, to the slow process of repetition by which light was introduced to the world and the importance of ameliorating its effects. Rather than basking in the glare of reason, Māori mythology emphasises a need for respite, change and complement. Carl Mika, a Māori academic, notes that 'we are entangled in, and appropriated by, a colonial fixation on clarity, visibility and enlightenment'. He recognises a need to throw off this 'obsession', calling instead for us to 'treat obscurity, invisibility and endarkenment benevolently'. Enlightenment – endarkenment.

Why strain for one and ignore the other? Is it really so terrible to let darkness fall? Mika encourages us to step out of the light and into the dark. What do we have to lose?

Unease, fear, concealment, madness, even evil – darkness is too often a thing to be shunned, avoided, broken. But what about the softer side of it? Dad's illness is terrifying. It makes ideas of the dark newly strange and sinister. It makes the psychological reality of darkness immediately physical, and the sensory floundering deeply psychological. It tests and complicates everything I have until now considered darkness to be. But I want to take with me Mika's notion of benevolence. I don't want to lose my appreciation of the wonder and beauty of darkness, the relief it can provide and the comfort; I'm interested in challenging the prioritisation of the clear, the visible and the light. Dad is being, in Mika's phrase, endarkened. There is nothing to be done. But because of this, and now more than ever, as the moon wanes and the natural cycle closes in, it seems important to fully celebrate the dark.

And so I'm beginning to think about death.

CHAPTER 4

# Waning

All his life my dad has loved to watch football. Now the scurry and noise confuse and annoy him. But on Cup Final day, the teams spill onto the TV in the corner of the lounge and he's distracted momentarily by the flicks of colour as the camera pans across the Wembley stands. It's almost kick off. The crowd sings. My dad weeps.

> Abide with me, fast falls the eventide.
> The darkness deepens ...

For almost a hundred years, since 1927, this Victorian hymn of mourning and death has opened the FA Cup Final; since 1929 it has also been sung at every Rugby League Challenge Cup Final. On mid-afternoon sunlit pitches, days of sporting achievement throughout my dad's lifetime have begun with a moment of solemnity, a nod to the transience of human endeavour – and an evocation of darkness. Leading us from the brightness of day into evening and on into night, 'Abide with Me' describes a loosening of the light: 'life's little day', we sing, gives way to 'gloom'. The day is short, 'little'; the gloom is vast. It's a hymn that reminds us of Bede's sparrow, flitting through the lit barn, drawing on the Christian metaphors we've already seen in the hope of 'heaven's morning' breaking and, with it, the prospect of resurrection. But at its heart, this is a hymn shrouded in darkness. Written by the dying Anglican minister, Henry Lyte, and usually sung (as it is on

Cup Final day) to a tune called 'Eventide', it's a plea and a prayer in the face of the encroaching, undeniable dark.

I watch dad watching the military band play the final bars. It's a rainy day and all the lights are on. The telly fidgets bright. But this is a room of shadows, and my dad peers, as though he needs more light.

Which brings us back, I suppose, to Dylan Thomas. Not, this time, the 'slow black, sloe-black' dark of Llareggub in *Under Milk Wood* but his famous exhortation to 'Rage, rage against the dying of the light'. The phrase echoes through his fiery poem, 'Do Not Go Gentle Into That Good Night'. Addressed to an ailing father, 'Do Not Go Gentle' is again an entreaty in the face of the dark, a summoning of courage which takes on the resonance and rhythm of the pulpit; like 'Abide with Me' – and for all its talk of light – it's a hymn to the power of darkness. It comes to mind as I watch my dad, floundering more often than raging, baffled by the dying of the light more often than angered by it: the more lights dad turns on, the more he finds himself mired in shadow. But he's clearly not alone. 'Do Not Go Gentle' was apparently one of the most popular poems of the twentieth century, commonly topping polls, which seems to evidence both our vehement resistance to the advance of the dark and a continuing search for ways of acknowledging and expressing the process of dying.

What we understand of the dark brings us closer to what we understand of death.

This connection proves inescapable in language, which has long used dark as a synonym for death, even beyond an obvious religious context. In the past, the darkest passage of the night, between midnight and dawn, was referred to as 'dead time' or 'the dead hour'. In *Hamlet,* the ghost appears to the guards 'in the dead vast and middle of the night'. Shakespeare uses a similar metaphor in his long poem, *The Rape of Lucrece*, when he summons the eerie menace of 'the dead of night' – a phrase we still use today – when even the stars are absent and the only sound is the 'death-boding cries' of the hunting wolves. One particularly striking seventeenth-century prayer, called on God to intervene 'now that the darke has come, which is a sign of horror, death and woe', pointing out that laying down in bed to sleep was 'an image of the grave wherein my body after this life is like to rest'. As Dylan Thomas notes in another, lesser-known poem, 'A Refusal to Mourn the Death, by Fire, of a Child in London', death is a state akin to the 'Fathering and all humbling darkness', a state potent with, and yet ultimately resistant to, endings and beginnings, loss and promise, even time itself.

For centuries, imagery of the dark has been the go-to means of edging towards an understanding and description of death; the prevalence of metaphorical language linking darkness with death shows just how deep-seated and instinctive are the connections we make between the two. Such associations begin early: when we're young, our natural curiosity about death commonly expresses itself in a fascination with the dark – the witches, ghouls and monsters which lurk there and the strange, disquieting feeling of being alone in darkness. The sense of dissolution and dislocation, the strangeness, the fleeting feeling of incorporeality, the suspension of material cues and sensory markers which we recognise in the dark, all help nudge us towards some kind of estimation of what death might be like. This is how the German writer, Antje Rávic Strubel, describes the ways she tried to imagine death as a child, for example:

As a child, I was maybe four or five years old, I often tried to imagine being dead. I lay completely still in my bed, eyes wide open, staring into the dark room. I believed that the less I moved, the more I was dead. The night helped me. The darkness and my inertness merged into the dense black that I believed non-being consisted in. Neither the pain of having to part, nor the fear of the pain of dying played a role in those early years; when the reality of death was so distant; but rather the question of how something that had existed could suddenly cease to exist. I stared into the dark room and couldn't conceive that beyond this darkness there was more darkness, an infinite darkness, an endless dark vastness that would exist in the space where I no longer was; yet at the same time it couldn't possibly be dark anymore because I would no longer be there to perceive it. It was like being sucked into a tube that grew wider as I got smaller, until the borders of it disappeared and expanded into infinity, and I tried desperately to recognise something in the dark, the outlines of my desk, the moonlight through the crack between the window and blinds. Everything dissolved.

Darkness and more darkness, 'an infinite darkness' without shape or borders, all achieved by staring into the inert 'dense black' from the comfort of bed. Strubel's childhood experiment seems simple enough – stare into the dark and allow the world to dissolve – but it's remarkably effective in loosening our connection to our physical self and slipping into the dark.

So are we closer to death in the dark?

Strubel's account of 'being sucked into a tube' as she lies still and quiet trying to imagine death chimes with the descriptions given by people who have gone beyond childhood imaginings and – as a result of accident or illness – have slipped between life and death in what are known as 'near-death experiences'. These experiences can be terrifying, comforting, life-changing, dramatic or quiet, but in all of the stories, darkness comes alive.

In some accounts, these near-death experiences hinge on the darkness giving way to light. A transit through the dark often signals the slide from life into death, a changing state of consciousness beyond which is the promise of light returning. People often talk about walking in the dark of a cave or a tunnel, for example, or feeling as though they're moving through a passage or climbing a dark staircase. Even for those without any kind of religious belief, there's often a powerful beam of light or a bright space in the distance, a glimpse of some kind of paradisal promise which acts as a magnetic draw. The writer, Ernest Hemingway, wrote home about a similar experience after being hit and seriously injured by an exploding shell in Italy during the First World War, when he was nineteen years old. He later drew on his experience as the basis for a short story, 'The Snows of Kilimanjaro' (1936), which uses a plane journey into a storm to try to capture the essence of the classic near-death plunge into darkness before an emergence into light, infused with a feeling of peace and joy, a contentment at arrival:

> Then they began to climb and they were going to the East it seemed, and then it darkened and they were in a storm, the rain so thick it seemed like flying through a waterfall, and then they were out ... all he could see, as wide as all the world, great, high, and unbelievably white in the sun, was the square top of Kilimanjaro. And then he knew that there was where he was going.

'And then it darkened': Hemingway's evocation of dying matches many others. Dark closing in; dark upon us. Whether real or not – and scientists are still not sure how to explain experiences such as the one Hemingway remembered – near-death narratives are testament to the grip of the dark on our concept of death.

In some of the most striking of such testaments, light is absent altogether. Death-dark becomes a fully immersive state, dying a complete and overwhelming sensory submersion into darkness. Some people describe a huge terrifying emptiness that disorientates and isolates, where time is absent or distorted and there is nothing but an endless floundering in a void: 'I felt the aloneness, the emptiness of space, the vastness of the universe ... I was being drawn into this dark abyss, or tunnel, or void ... I was not aware of my body as I know it ... I suddenly was surrounded by total blackness, floating in nothing but black space, with no up, no down, left, or right ...' More commonly, though, the dark is benign. 'I went into a dark place with nothing around me, but I wasn't scared,' one person reported. Another, a Cypriot woman, described 'floating towards a dark, but not frightening, curtain-like area'. An American man, who found himself trapped in the churning rapids below a waterfall and almost drowned, recounted how the dark felt full of possibility, his mind racing through scenes from life as he floundered in the water until he reached a state of quiet anticipation: 'there was only darkness and a feeling of a short pause, like something was about to happen.' In such moments, darkness is a solace, a thing of potential. Many of the people who have lived through this kind of near-death episode talk afterwards about losing their fear of death, and at the same time feeling a new craving for dark spaces and the immersive consolation of the swaddling dark.

The act of dying, it seems, draws us into the dark, a dark without form or limit, a dark of new experience, a lull, a promise, unbodied and timeless. We'll find out, one day, whether this is true; for now we can only read these accounts and wonder.

★⁺₊★ ☾★⁺₊★

Or we can sleep. The one other state that we associate with death and the dark, that seems to hold us poised in stasis, in a new dimension, suspending us even on the brightest of days in an internal darkness that shifts and changes, overwhelms, alarms and relieves, each time we experience it. Like darkness and death, sleep is egalitarian: it comes to us all, and elides differences in rank, education or wealth.

Shakespeare's warrior king, Macbeth, a man consumed with guilty insomnia, describes sleep as 'the death of each day's life'. This sense of sleep as a precursor to death – some kind of nightly dress rehearsal for the big event – is a common motif: Homer called sleep and death 'twin brothers', and Alexander Pope described them as 'two twins of winged race'. Franz Kafka wrote about resisting sleep as a way of warding off death, suggesting his insomnia was caused by a 'great fear' of dying. More recently, in his novel *Lisey's Story* (2006), Stephen King describes how Lisey is 'lost in the land of sleep … afraid that it is also a preview of death'. The porous boundaries between the two states have also been explored by many writers struggling to find ways of grasping some kind of understanding of death: Edgar Allan Poe called sleep 'those little slices of death'; in his poem 'On Death' (1814), written while caring for his dying brother, John Keats asked 'Can death be sleep when life is but a dream?' Like the dark, sleep is also, of course, a common metaphor for death – think of euphemisms such as 'the big sleep' or 'going to your rest', or the seventeenth-century prayer we've already seen which expressed lying asleep in bed as a foreshadowing of lying in your coffin.

Sleep, death, darkness – they're all slippery and strange; they all dodge and shimmy to evade our efforts to describe them. How do you explain what it's like to fall asleep? How do you pin down that moment when you drift from some sort of waking to some sort of sleeping? Floating, sinking, sliding: much of the language of falling asleep recalls the ways people describe near-death experiences. And both return to the kind of language we've come to associate with the dark. Darkness falls; we fall asleep.

Aristotle spent some time trying to unpick the nature of sleep and sleeplessness, and in his efforts to capture the moment of falling asleep, he focussed on the senses. We know when we're awake, he claimed, because we sense movement, either externally or in our own consciousness. We know when we're asleep because our sense of touch – which he identified as the primary sense, 'our common and controlling organ of sense perception' – becomes incapacitated. Later, Freud claimed that in falling asleep we withdraw from the world, becoming suspended in a kind of non-state; the twentieth-century French philosopher, Jean-Luc Nancy, describes the process of falling asleep as a process of disappearing: 'I fall asleep and at the same time I vanish.' Sensory deprivation, hovering in a void, the erasing of self, all of these ideas resonate with our notions of death, and all of them chime with our experience of the dark. In the dark, as in sleep, we're thrown into an absence, we're there and not there; in the dark we come closest to Nancy's moment of being and not-being, of falling towards the vanishing point.

In our efforts to sleep – and there's a burgeoning body of contemporary writing reflecting the modern preoccupation with insomnia – we generally embrace darkness. We close curtains and blinds, turn off lights, perhaps even wear eye masks or put up black-out linings. We welcome the dark as a good thing. As we've already seen, we're biologically programmed to need darkness to sleep most effectively. And in the animal world, of course, there are a number of species which fully embrace the dark to enter hibernation. Hedgehogs, bears, bats, dormice and bumblebees are among the creatures which actively seek out darkness in which to pass the winter: bears famously curl up in caves, bumblebees and mice burrow into the earth. The protection of the dark allows their bodies to shut down into a state of torpor that slows heart rate, breathing and energy consumption – and so allows them to survive periods when food is scarce. In the past, it's been claimed that peasant families similarly withdrew through the long dark months of winter, saving food, heating and energy by nesting in bed together in a kind of hibernation. An article in the *British Medical Journal* in 1900, for example, explained how families in rural Russia survived the miserable drag of the winter months:

> At the first fall of snow the whole family gathers round the stove, lies down, ceases to wrestle with the problems of human existence, and quietly goes to sleep. Once a day every one wakes up to eat a piece of hard bread, of which an amount sufficient to last six months has providently been baked in the previous autumn. When the bread has been washed down with a draught of water, everyone goes to sleep again. The members of the family take it in turn to watch and keep the fire alight. After six months of this reposeful existence the family wakes up, shakes itself, goes out to see if the grass is growing, and by-and-by sets to work at summer tasks.

Some modern historians have cast doubt on accounts of such bucolic simplicity – and six months of sleep does feel like a long time – but it's not impossible that in the harshest conditions, some kind of 'hibernation' response was triggered that allowed rural families to survive. More recently, a similar reaction was found in scientists holed up in the Antarctic research stations during the winter who entered a state of 'psychological hibernation' as a reaction to the dark and cold. They recorded feelings of indifference and emotional flatness that were thought to act as a protective mechanism against the chronic stress of mid-winter conditions, recalling the similar retreat recorded in more colourful terms by the *British Medical Journal*.

Such periods of listlessness and inactivity are clearly a way of making it through the hardest months rather than genuine metabolic hibernation. We've all felt the urge to snuggle up and sleep during a period of bitter weather, the dark days stimulating a need to withdraw and keep warm. But in fact, while we often think of this as a hibernation response, more recent research suggests that hibernation is actually not so much about curling up to sleep as about sleep deprivation. It's been found that when animals emerge from hibernation in the dark, their brain-wave patterns are similar to when they've been sleep deprived. In addition, most hibernating animals regularly come out of their torpor state specifically to sleep. This suggests that the period of hibernation is actually a time when the body is unable to properly sleep, a state of greatly reduced consciousness without movement or sensory activity, without thought or feeling. While we might like to think of it as a kind of sleep, it's much more like hovering close to death.

While plenty of research has looked at how animals enter into, and maintain, the process of hibernation, there are few conclusive answers: on the one hand, it's possible that it's triggered by changes in individual cells of the body at a molecular level; on the other, it may involve signals from the brain and nervous system, or input from hormones. Most studies suggest that fasting and temperature play a part. It's generally agreed, however, that darkness is essential. In hibernation, the dark acts to shut us down. By contrast, when we're sleeping, it's in the dark that our dreams take shape. The ancient Sanskrit medical text, *Sushruta Samhita,* explains how sleep and dreams are dependent on the quality of the *tamas* (darkness). In the *Samhita*, the turmoil of a dreamy sleep is shown to prevent prayer and learning, bad dreams being sent by hostile spirits and friends appearing to deliver threatening messages. A chapter in the eleventh or twelfth century Jain text, *Mahānisīhasutta,* is given over to listing the atonements required from monks who have forgotten to utter the relevant mantras for keeping the dangers of darkness – mostly wild beasts and demons – at bay while they sleep and dream. In Edgar Allen Poe's poem, 'The Raven', the narrator is shown 'deep into that darkness peering ... dreaming dreams no mortal ever dared to dream before.' From the earliest cultures to modern times, the power of dreams has been a fascination nursed in the dark.

It's usually in the darkest hours that our dream life is most vivid. When sleep cycles were first discovered in the early twentieth century, by tracing the electrical patterns of brain activity, it quickly became apparent that there were the two types of sleep most of us are now familiar with: non-rapid-eye movement (NREM) and rapid-eye-movement (REM). Exactly why we slip between the two types still isn't understood, but it's agreed that REM accounts for between 20 and 25 per cent of a normal night's sleep, with REM sleep increasing as the night progresses until reaching its peak in the hours before we wake. During REM sleep our brains show increased activity in the regions responsible for visual, motor, emotional and autobiographical memory, and a decrease in the areas that govern rational thought – so REM sleep is responsible for our most lively dreaming, drawing us into a typically vivid and bizarre dreamscape, weaving elements of our waking life into an incoherent, fantastical but (to us) hyper-real narrative. Research has found that these dreams can have a positive effect on the brain, improving memory and mood, helping us read emotions and judge facial expression in our waking life, lowering fear responses to external stimuli and helping protect against conditions such as post-traumatic stress disorder, possibly through a recalibration of the stress hormone, norepinephrine, which is reset during REM sleep. But for all these things to happen, we need darkness. Too much light reduces REM duration, or prevents us from moving into the REM stage at all. Deprived of darkness, our dreams are vapid, too slight to restore or refresh, too insipid to matter. The weirdness of dreaming is the magic of the deepest dark.

The moon is waning, vanishing. Its cycle is drawing to a close and, gradually, it rises later in the night, until it slips quietly into daylight. It becomes a daytime moon again, visible in the morning, but slight and unspectacular. Unanchored from the dark, it grows smaller and smaller, losing itself in the light.

The evenings, moonless, resound with dark again.

Like the moon, my dad is waning, vanishing. The black bud of his dementia flowers darkness inside him; behind him, he trails shadow. He has moved into a care home, a red-brick new-build by a dual carriageway. It is a calm, lit place of carefully maintained sameness, timeless and placeless, safe. As I leave after my visit, the autumn night is drawing in. A stream of traffic trembles white and red at the end of the drive, but behind the tall trees edging the car park, the dusk is dense, blackening. I shimmy through the trees and into a small wood. I walk for a while. Towards the far boundary, where the shadows break away into open fields, a tabby cat sits and stares at me, its eyes bright. I turn at the edge of the woods to make my way back to the car. I can see the lights from the care home blinking through the trees as the branches ruffle in the wind and I think about my dad, pacing the corridors under the brilliant, unchanging strip lights.

The cat slinks silently under a fence and disappears in the deep pool of shadow beyond.

Cats' eyes. Piercing and strange, they've been noted since Ancient Egyptian times for their otherworldly glow in the dark. As a nocturnal hunter, the cat has especially large eyes, made for stalking in darkness, with pupils that open particularly wide. Some other nocturnal animals take these outsized eyes to extremes: the tarsier, for example, a South East Asian primate, has two huge beady eyes, each one the size of its brain, allowing it to hunt tiny insects even in the darkest conditions. But animals with good night vision – owls, tree frogs, racoons, some snakes, cuttlefish, foxes, geckoes, cockroaches – don't only rely on the size of their eyes. In owls, for example, the shape, too, is different, tubular rather than spherical, allowing a lot of light to be drawn into the retina. In nocturnal predators, the nuclei of the rod cells are densest in the centre of the eye – in contrast to us, where the densest areas are around the edges – effectively focussing any light into one area, while most also have a mirror-like layer behind the retina, called a tapetum lucidum. This helps them exploit even tiny amounts of light by bouncing it back, giving the retinal cells a second chance at picking it up. It's this shiny reflector that makes the eyes glow in the dark.

The darkest environments on earth are home to some of the strangest see-in-the-dark creatures. Cave systems can stretch for hundreds of miles and reach depths of over five thousand feet, well beyond the reach of natural light. Nonetheless, some highly specialised creatures feel at home there and manage to find a way to navigate the absolute dark. Some of these animals use infrared vision. The cave boa, for example, is a snake which hangs from the ceilings of caves in South America. It can detect the tiny spots of heat from passing bats allowing them to hunt effectively in complete darkness. Others have evolved to do without sight altogether, making up for the lack of vision by clever use of other senses. Cave fish are exceptionally sensitive to pressure and vibration and have a highly developed sense of smell that they use to guide them. Since there's no day and night in a cave, they have also evolved to do without sleep, constantly swimming in the search for food. The Mexican cave fish called the tetra has genetically diverged into two forms – the one which lives above ground has eyes and a coloured skin; the form which swims permanently in the dark has sunken cysts instead of eyes, and has lost all pigmentation. Cave-dwelling salamanders also manage without eyes, having developed electroreceptors that enable them to sense approaching prey, and to avoid predators.

In the absence of reliable sight, other senses allow animals to live and hunt effectively in the darkest conditions. Owls and large cats have highly developed hearing and specially adapted ears. Many nocturnal creatures have an excellent sense of smell, communicating through scent-marking; snakes use taste to locate prey. Touch can also be very sensitive, allowing animals to pick up vibrations in the dark. Hades centipedes, found in only three caves in the Helebit mountains in Croatia, use their antennae to feel for prey. Colourless and blind, remipedes, the only known venomous crustacean, have chemosensory hairs on their antennae which can detect chemical molecules particular to their prey. Harvest mice, rats, shrews and voles make expert use of their whiskers: dormice flick their whiskers ten times a second to help them understand the space around them.

The experience of darkness makes us pause, too, to reconsider the primacy of sight. For a very long time, the Western model of understanding – science, politics, social structures – has been largely built on what we can see, and expressed through visual language. Aristotle raised the sense of sight 'above all others' because it 'makes us know and brings to light many differences between things'. Plato described sight as 'the source of the great-est benefit'. By the Renaissance, the eye was firmly established as the organ of most power, speed and accuracy, providing the most direct knowledge of the world. The sixteenth-century eye-surgeon Georg Bartisch summed up general opinion when he described the eye as 'the most necessary, noblest, clearest and most subtle member above all others'. As printing and reading took a firm hold on knowledge, learning was considered impossible without sight; astronomy, anatomy, mathematics, fashion, painting, gar-dens, theatre and religious worship all relied on vision, display and spectacle. Even the inward eye was important – memory and phi-losophy were dependent on being able to see and interpret images in the mind. Vision, of course, is (and was) often unreliable and deceptive. The evidence of the eye is not always what is seems. Even in Renaissance Europe there was already disagreement about the nature and trustworthiness of perception; more recently commentators have begun to unpick our dependence on sight and our assumptions that things visible are therefore intelligible. In the twentieth century, writers and psychoanalysts such as Michel Foucault and Jacques Lacan challenged the historical confidence in the eye; the emergence of virtual reality, CGI, and deep fakes have tested our belief in visual material much further. Even so, we've grown used to relying on our ways of seeing the world.

The blindness of the dark can be a revelation. For those of us who can usually see, there's a strangeness that forces us to experience our world differently, an ease that allows us to ignore the clamour of the visual. We comprehend our environment in a new way. We're an unaccustomed version of ourselves, our senses stretched and rearranged. Looking into the dark, what it is we see?

Think about walking by a freshly cut field at night. Unable to *see* the field, our sense of it intensifies – the complex scents of hay and earth are utterly distinctive; the warm air drifting up from the drying grass tingles on the skin. We can't know the field's shape or scope but in the dark, it's a smudge of tones, a commotion of instincts; the shuffle of the wind through stalks perhaps, or the metallic rub of crickets, the strange click and rattle of the nightjar. In darkness, stripped of sight, the field becomes a multi-sensory adventure. It draws us out into the unseen world, but also throws us back on ourselves. Juhani Pallasmaa writes about the 'acoustic intimacy' that occurs at times like this when the non-visual senses are engaged, evoking an experience of interiority where the boundaries between inside and outside become distorted and porous: 'the space traced by the ear in darkness,' he notices, 'becomes a cavity sculpted directly in the interior of the mind … sight isolates, whereas sound incorporates.' He describes how the ear – listening perhaps to the drip of water in a dark ruin – has the ability to shape sound through imagination, to 'carve a volume into the void of darkness'. In his moving account of the onset of blindness, John Hull describes a similar intensity of experience that changes the way he thinks of himself and his body in the encompassing dark. He notes that darkness can 'turn off' the visible world, but the acoustic and tactile worlds can't be shut out. Instead they offer difference, an alternative way of understanding the physical environment: 'I am developing the art of gazing with my hands,' he writes, and in listening to the peal of church bells he describes how 'the ground seemed to be trembling, and the very air was heavy and springy with the reverberations ... To me, the very air I was breathing was bell-shaped'.

While the dark of the past is forever lost to us, such a sensory experience gives us the best chance of recovering an impression of what it might, perhaps, have been like. Our impressions of darkness are never static, changing with place and weather, with mood and opportunity, with time. Tonight's dark won't be the same as last night's; the dark of a deep cave feels very different to the dark of an unlit room. But if we resist the urge to rely on sight, we might find a way of reaching back. The dark of the Middle Ages or the Renaissance, for example, is likely to have been a much smellier affair than we're used to. In the quiet of the night, pungent smells would have bloomed. If you weren't lucky enough to afford a bed, you'd be hunkering down to sleep with the entire family in a tumble of hay or straw, sacking, old cloth, wool, horsehair, feathers or even pea-pods, the dark coming alive with the outdoor scent of herbs crushed on the earth floor or matting, cattle or pigs perhaps shuffling in their stalls underneath the room or just the other side of the bed, the communal chamber pot or night-stool stashed beneath. Closed up and shuttered for the night, the house would have stored all the scents of the day from working, cooking, cleaning, tending to animals and children, treating disease or welcoming guests, a tangle of blood and urine, wine and meat, sweat and sex, smoke and damp. Your body and breath would smell too, of course, as would the person snuggled alongside you or snoring at your feet. Even among richer classes, a commitment to personal hygiene remained slight until well into the nineteenth century. The French historian, Alain Corbin, memorably described the ordinary masses of post-Revolutionary France, for example, as 'putrid... stinking like death, like sin.'

Fragrant and foul, the dark of the past was also a clatter and hum, a soundscape mapped by sensitive ears. In her essay 'The Gender of Sound', Canadian poet and classicist Anne Carson reminds us that we've long relied on non-visual clues to assess friends and strangers alike: 'It is in large part according to the sounds people make that we judge them sane or insane, male or female, good, evil, trustworthy, depressive, marriageable, moribund, likely or unlikely to make war on us, little better than animals, inspired by God.' When night-time conditions were frequently and genuinely dark, when rooms were shadowy and often unlit, when recourse to lamps and candles was reserved for work or special occasion, how much more did we rely on sound to guide us? But as Carson notes, such judgements, unmediated by sight, 'happen fast and can be brutal'. An old Scottish proverb reminds us to be on our guard against the ruthless assessment of night-time utterance; deeds that might be spotted in daylight can just as easily be overheard at night: 'the day has eyes, the night has ears.' In the still darkness, sound travels; whispers resonate. Yorkshire dialect also historically draws attention to this: the verb 'to dark' was used to describe the act of listening in unperceived. Eavesdropping is easier in the dark. And Shakespeare's Hermia, in A Midsummer Night's Dream, again reminds us how the ear takes over the power of the eye once darkness falls: 'Dark night, that from the eye his function takes/ The ear more quick of apprehension makes/ wherein it doth impair the seeing sense/ It pays the hearing double recompense.' Hearing quickened, catching the subtle and multi-layered sounds of the dark: church bells and ticking clocks, the crackling of fires and beams, the turn of a page, the rustle of animals and bedfellows.

Many of the whispers and murmurs the other side of silence were domestic. The long mid-eighteenth-century poem known as 'Night Thoughts', by Edward Young, delights in the calm of the dark when the 'tumultuous life' of the day can be set aside, and the thoughtful narrator can settle down at home and turn his mind to serious contemplation of matters of life, death and immortality:

> What awful joy! what mental liberty!
> I am not pent in darkness; rather say
> (If not too bold) in darkness I'm embower'd.
> Delightful gloom! the clustering thoughts around
> Spontaneous rise, and blossom in the shade

For Young, the darkness offered freedom and possibility, a time when he felt empowered and thoughts could 'blossom' spontaneously in a way that was usually prevented by the demands of daylight activity. Darkness offered the opportunity for rest. The toil of the day could be forgotten for a while and the body recharged. Just as importantly, it freed the mind from the cares of the everyday. This belief that the dark allows for clearer thinking, for spirituality and the release of the more cerebral side of our natures, was nurtured by many religions which urged their followers to use the quiet night hours at home to bring themselves closer to the divine. The early Christian church, for example, emphasised the importance of prayer at night: night vigils are known to have been held from the fourth century; in the early fifth century, Jerome strongly recommended the practice of nocturnal prayer, urging his followers to rise two or even three times a night to sit quietly in the dark; later, in a series of meditations known as the 'Spiritual Exercises', Ignatius of Loyola, the sixteenth-century founder of the Jesuit movement, recommended spending at least one period of prayer in the middle of the night.

At a time when the customs and rules of ecclesiastical practice were so influential, this led to an ingrained respect for the dark. Darkness was not only safe, but sacred. By the Middle Ages, all the processes of civil life were suspended with the onset of night: courts and councils, shops and markets all had to fall quiet at dusk. This was not only about protecting townsfolk from having to venture abroad, but also about protecting the unique quality of darkness. Since God was considered to favour the dark, dwelling in the infinite and unknowable that it represented, the Catholic Church carefully guarded night as a time for meditation and prayer, both personal and communal, in homes and in monasteries. From its rise in the sixteenth century, Protestant practice also particularly emphasised the importance of keeping the hours of darkness quiet and for devout reflection. There was a shared belief that darkness was an essential element of the divine plan, and should not be meddled with. Even by the eighteenth century, the Catholic Pope Gregory XVI was forbidding streetlights in case they prompted people into ungodly acts (though they had been shown to reduce crime), and when the Calvinist authorities in Geneva insisted on preserving the city's darkness to encourage prayer, the enlightenment philosopher Jean-Jacques Rousseau is widely reported to have commented (perhaps ironically) that 'God does not agree with the use of lanterns'.

But for many, Young's 'calm intercourse' was considered a terrible waste of the dark. While there were plenty of devout people
who set aside the time for prayer and contemplation, there were
plenty more of a less studious turn of mind for whom darkness
was much more animated, carnal and raucous. In popular culture, the dark became a byword for licentiousness and disorder,
thrumming with rebellious activity and noise: 'O thou the silent
darknesse of the night, arme me with desperate courage and
contempt ...' wrote the playwright Thomas Goffe in *The Raging
Turk* of 1631, 'to drink and swagger, and spurne at all the powers
of either world.' Under cover of darkness, the earthly and divine –
'the powers of either world' – could be 'spurned', disputed, or
at the very least, ignored for a few hours. Hierarchies were less
evident, the norms of respectable society obscured, and, out of
sight of their employers, workers could reclaim some autonomy.
Manual labourers, servants and slaves traditionally used the hours
of darkness to arrange romantic meetings, dances or drinking
sessions. While the ruling classes dazzled with the light of candles
and fireworks to make a spectacle of their authority, as we've seen,
those lower down the scale often found their dominion in the
dark. In *The Mayor of Casterbridge* (1886), Thomas Hardy – who
revels in the crusty characters of everyday rural life – notes that
'darkness endows the small and ordinary ones among mankind
with poetical power'.

Even when work still remained to be done, as it often did,
darkness allowed for a change of tone, a relaxing of the rhythms
of the day. While daytime tasks were frequently exhaustingly
physical – whether in the fields or in the factory – the night at
least allowed for sitting down, and shared hearths were perfect for gossip. There are numerous records of people meeting
regularly to spin, sew, clean hemp or knit, tasks which, for the

skilful, required little light and could be done while chatting, telling stories, or singing. These gatherings often took place in stables, workshops or barns, or even outside, and had their own names across Europe usually linked to a sense of vigil: *veillées* in France, for example, *veglia* in Italy and *kvöldvaka* in Iceland. Not only did this work often provide important extra income but it provided a way for women, in particular, to come together socially. Claiming the dark for community, swapping news, offering support and advice, and sharing workloads, these meetings meant the hours of darkness were productive and treasured. Jean-Jacques de Boisseau, a French artist interested in recording peasant life, depicted one of these gatherings in his painting, *Evening in the Village* (1800). It shows a group of adults and children clustered in a stone farmhouse or barn, the light from a fire on hands and faces. Babies and pets are sleeping on laps; a pipe is being shared, a private conversation takes place in a darkened recess, suggesting the sharing of secrets. Several families have come together. The characters are happy and relaxed, smiling. Boisseau's scene is companionable and comfortable, the dark a place of simple tasks and friendships. In the sturdy shadows of the old room, lives unfold.

I'm on a beach not far from home. It's early in the morning, and the sun's not due up for a while yet. The air's chill for November and I'm huddled in my coat, my hood pulled up and my hands stuffed in my pockets; in front of me the sea churns and spits. From this small bay I look west – I can see the waning moon hanging faintly yellow, still quite high in the sky, giving only a dim light. It's enough to make out the slight shimmer of the sea and the blotted shine of wet pebbles, but not much more. The beach is mostly sound: the churn of the waves and rattle of stones, the wind snapping at my coat and in the branches of the low trees clustering around a sharp-sided gully, the rush of water through the same gully from the hills behind.

I sit and wait for the night to pass. The moon sinks slowly, smudging the sky and the sea as it sets. Behind me on the beach, I cast the faintest of shadows in the fading moonlight, a hint of dark upon dark, as though I'm not quite corporeal. The shadow feels more solid than I do, even though it's barely there. It's possible that when I move on, it will remain until the tide comes in to wash it away.

The night-time shadow is a beautiful thing, ephemeral, haunting. I wonder about how it fits into the dark, where one ends and the other begins. Shadow is the nearly dark, a recollection of both light and darkness, belonging to both, requiring and elucidating both. In 1767, the French surgeon and scientist, Claude-Nicolas Le Cat, confidently declared that shadows were 'holes in the light', a thing missing, lacking, but I'm not making a hole, sitting here on the beach, am I? It's more like I'm adding to darkness, texturing or layering it. My shadow has its own body and presence. It sticks with me, the itinerant dark.

In the late fifteenth century, Leonardo da Vinci explored the effect of shadow on shapes, especially the shapes we find in architecture. He found subtle distinctions in the kinds of shadow made by different features, finally defining three distinct types which he called attached shadow, shading and cast shadow. The attached shadow, he explained, falls on the body itself – like a cantilever roof causing a shadow on the façade of a building. The second type, shading, is caused by bright and dark contrasts and by form: so a sphere will create a shadow below it, for example. The third, cast shadow, is the type of shadow we get when we stand in front of a source of light, the shadow we see on the ground on a sunny day, or the kind of shade produced by a tall house on a local street. It's these various types of shadow that many artists use to model figures and create the illusion that a scene is real. Japanese and Asian art, by contrast, traditionally avoids the use of shadow – we instantly read these works in a different way, as resolutely two-dimensional, with a flatness and graphic simplicity that is evocative of place and period.

But how many of us think about shadows? Unless, like Leonardo, we're studying built form or shading drawings to create heft and volume – or unless we're seeking a cool spot on a hot day – most of us don't notice shadows at all. We use them as spatial cues, instinctively helping us make out the shape and scale of an object (whether its convex or concave, for example, whether it's attached to the ground or floating, moving or still) but we don't *take notice* of the shadow itself. We normally see right through it. Our eyes hardly take in the smear of subtle dark. When you walk through a shadow, nothing happens. It doesn't cling to us, like a web. It isn't solid; it makes no noise. Like full-on darkness, it causes no physical impact; it has no resistance, no mass. Shadows can seem like nothing. Perhaps this is what led Le Cat to think of them as holes.

But of course when we look for them, shadows are everywhere. These whispers of the dark are upon us even – especially – in the brightest light.

Could we exist without shadows?

How do you think of them, those greyish blobs that shapes leave lying around, a kind of light litter? Do you think of them at all?

It turns out that mice are particularly alert to shadow. They use a specific neural pathway that allows them to detect even the dimmest of shadows in almost complete darkness, the merest blot of dark on dark, right at the limit of what's possible to see. But shadows are deeply entrenched in human optics, too – we share an almost identical neural pathway with mice, enabling an exquisite sensitivity to shadow. Our eyes developed around 530 million years ago from a simple skin patch that became photosensitive – that is, it became sensitive to light and dark. And an early function of this evolving eye was shadow detection. This allowed us to distinguish form, size and distance, so we could successfully navigate, hunt, and avoid predators. Shadow was important, a matter of life and death. We were acutely aware of the relationship between a shadow and the thing which was casting it; we understood the 'meaning' of shadow. These days we prefer to ignore it. But when we look at it closely, we see that shadow, like the dark it mimics, has character and function. Shadows can cause us to reconsider what we think we know about how we perceive the world around us.

Let's go back to Leonardo's studies and the representation of shapes on paper, in paint or on screen. Because here shadows are not as simple as they seem. They're tricksters, illusionists. When we add shadow to an object – in a sketch or photograph, on a computer – logic tells us that it should make the object appear lighter, since the shadow is darker by comparison. But actually it reverses our expectation: it acts to make the object appear darker. This is because our brain looks at the shadow and infers from it that the object is being spotlighted. In a clever shimmy, it then lowers its estimate of the object's intrinsic brightness, telling us that it's actually darker than it is. Shadows help remind us, then, that dark and light are not absolutes. They're negotiated, fickle – it depends how our brain decides to interpret them.

Colour, too, equivocates with shadow. If you look at a painting of shadow on snow, you'll notice that the painted shadows are not rendered in grey or black as you might expect, but in tones of blue. Claude Monet's *Lavacourt Under Snow* (1878–81), for example, shows a line of stone farm buildings at the edge of a winter field piled high in snow, and drenched in shades of blue; similarly, Van Gogh's *Snowy Landscape with Arles in the Background* (1888) marks out the shadows of footprints and drifts in pale, cool blue. This isn't just artistic licence. If you look closely at real shadows on snow, especially when the sun is low, you'll see that their darkness is a thing of unexpected colour. They are well and truly blue, like the blue dark of the Norwegian winter. This is caused by the scattering of light particles allowing shorter blue rays of light to reach the eye: the white snow adds no colour to the blue particles, which remain untainted, so in the dark shadows we glimpse the blue tinge.

The same effect – known as 'Rayleigh scattering' after the nineteenth-century British physicist and Nobel Prize winner, Lord Rayleigh – accounts for the fact that the sky appears blue on a fine day. And without the blue sky, shadows would seem eerily black, like fragments of pure darkness. When the Apollo 11 mission landed on the moon in 1969, these dark shadows took the astronauts by surprise, upsetting visual cues and making it difficult for the mind to grasp this bewildering new environment. 'Continually moving back and forth from sunlight to shadow should be avoided because it's going to cost you some time in perception ability,' noted Buzz Aldrin. In fact, despite the astronauts' initial impressions, moon shadows are not completely black. Sunlight reflected from the surface, and from the Earth, provide feeble sources of light that are just enough to mitigate absolute darkness, but photos show the shadows are much darker than any we're used to.

Back on earth, we piece together information from these scraps of darkness; we glean visual clues about the physical world, constructing what seems like a complete picture of places and things. But shadows keep moving, changing. They slide, bulge and shrink, and as they do so, they adjust the way we see things. They intimate time and juggle the complex physics of light and air. They give body to a flat world. 'Were it not for shadows,' wrote Jun'ichirō Tanizaki, 'there would be no beauty.'

And here's a fact I enjoy: if you stand in your garden at sunset, casting a long shadow across the lawn as the sun dips towards the horizon, the air inside that shadow weighs more than you do. (Air is heavy, and the elongated shadow accounts for a considerable volume of air.) But the shadow itself? That weighs nothing.

Shadows are the dark in daytime, creeping behind us as we walk, lurking to the side of us, anchoring darkness. They flit in and out with the sun, insubstantial, transitory. This strange status of shadows has made them such an evocative metaphor for our insubstantial lives. Shakespeare's Macbeth famously describes existence as 'a walking shadow, a poor player / That struts and frets his hour upon the stage / And then is heard no more'. The twelfth-century poem translated as *The Rubaiyat of Omar Khayyam* is even more effective in summoning the transience of our shadow lives:

> For in and out, above, about, below,
> 'Tis nothing but a Magic Shadow-show
> Play'd in a Box whose Candle is the Sun,
> Round which we Phantom Figures come and go.

Across most cultures, the comparison of life, or aspects of it, to shadow is commonplace while folklore, myth and legend not only use shadow as a metaphor for the brevity of existence, but also frequently propose the idea that shadows represent the life force of a person in some way, their essence or their soul. In the Canadian folk story, 'The Boy in the Land of Shadows', for example, the boy undertakes a long difficult journey in search of his dead sister, finding her in a garden where the last vestiges of the lost exist as shadows. He is given his own shadow by the voice of his sister who promises that the shadow will keep him safe and free of the darkness.

While in this story, the dead can only be seen as shadows, more often tradition dictates that ghosts and vampires – and all things dead – have no shadow at all. As our life force is spent, our shadows fade. When Peter Pan's shadow is pulled off and subsequently rolled up and tidied away in a drawer in J. M. Barrie's *Peter Pan* (1904), the boy is understandably lost without it. Barrie shows him 'shuddering', suspended in a fretful half-life, only becoming whole again when Wendy sews the shadow back on for him. In his unpublished jottings for *Peter Pan*, now known as 'Fairy notes', Barrie explores the intrinsic relationship between shadow and self that inspired the episode. He describes both the shadow and person experiencing the same physical and emotional traumas, even when they're separated. The shadow becomes a point of vulnerability, a way to attack or hurt the person to whom it belongs. Sketching out his ideas, he proposes a 'Girl suffering from want of her shadow – shadow also suffers, dwindles, &c,' or a 'shadow is quivering {therefore} original is suffering somewhere'. Finally, he muses, 'Suppose you cd hurt Peter by hurting his shadow, &c, (as in Indian fairy tale)'.

Shadows as live things; sentient shadows quivering, dwindling, hurting. What part of us do we see when we glimpse our shadows?

Barrie's notes are much bleaker in tone than the light-hearted *Peter Pan*. With their emphasis on suffering, they draw attention to the darker side of shadow as we see it in many traditions, bringing together the sense of an intrinsic life force with a recognition of our slippery natures. Hans Christian Andersen's disturbing 1847 story, 'The Shadow', for example, tells of a shadow that comes to life, travelling independently from its scholar 'owner' to investigate all the dark behaviours of humankind. 'Everything,' the shadow boasts. 'I saw and know everything.' Armed with this knowledge, the shadow manoeuvres into the affections of a princess, ultimately usurping the scholar's reality and claiming it's actually the scholar who is the shadow. In a final act of exchange, the shadow celebrates its marriage to the princess in the brilliance of the illuminated town while he throws the scholar into the dark and executes him. By the end of the nineteenth century, this idea of a shadow as a 'real' thing was also being explored by artists who started to consider shadow as somehow independent from the person casting it and capable of suggesting elements of personality that were otherwise difficult to grasp. In an echo of Hans Christian Andersen's shadow-without-a-body, Paul Gaugin went as far as recommending painting 'the shadow only of a person' rather than the figure. This 'original point of departure', he claimed, would capture the elusive 'strangeness' of a character.

In Andersen's story, the shadow wins out, ultimately living a glamourous independent life on its own terms, but more usually we're stuck with our shadows. They're impossible to throw off in anything other than a fairy tale. This idea of inescapable, even unwilling, attachment often mingles with the blackening effect of shadow to draw us back to the dark. The caution that our deeds

follow us like a shadow, unshakeable, and that our reputations cling
to us, crops up in a number of variations on popular sayings in many
cultural traditions: 'so the man, so his shadow', for example, or 'you
cannot outrun your shadow'. Agatha Christie was apparently particularly fond of this lingering sense of darkness, using it to set up
the plots of her crime novels; several of her characters – including
her famous detective Hercule Poirot – remark that 'old sins cast
long shadows' to explain how old deeds and grubby secrets catch
up with us. Christie was more or less contemporaneous with the
Swiss psychiatrist and psychoanalyst, Carl Jung, and may well have
been influenced by his work, which was deeply engaged in investigating the shadow side of our psyches. Jung considered that each
of us carries with us a 'shadow self', a complex mix of personal
and collective influences that results in elements of insight and
creativity, but also 'morally reprehensible tendencies'. The shadow,
Jung claimed, was a throwback to our animal origins buried deep in
our unconscious, 'that hidden, repressed, for the most part inferior
and guilt-laden personality whose ultimate ramifications reach
back into the realm of our animal ancestors ...' Jung proposed the
shadow as the dark side of our personalities, often deeply buried
but ultimately powerful, nurturing all the qualities and impulses
we would prefer to ignore or conceal.

You can't see an object by seeing its shadow. You can make a good
guess at what the object might be, but you have to look at the
thing itself to be sure; the shadow remains only an intimation.
On the other hand, when you look at a silhouette, you're looking
directly at the object itself, although in two dimensions instead
of three. While shadows are thrown out by an object, silhouettes

work by blocking a light source behind them so what we see is a black shape. We're looking only at surface, but in this case it's the 'real' surface, not the shadow. And like shadows, silhouettes can act as important visual cues: the eyes of nocturnal animals, for example, are particularly good at defining silhouettes as they hunt in the half-light. Like shadow, too, the silhouette is more than just a slither of the dark.

Silhouettes were made fashionable – and named after – Étienne de Silhouette, an eighteenth-century French politician and finance minister to Louis XV who spent his leisure hours cutting portraits out of black paper. There was a close link to shadow, in that silhouettes were sometimes made by tracing a shadow cast by a candle onto paper – early examples were also known as 'shades' or 'shadow portraits'. But as the idea took off, silhouettes often dispensed with the light source altogether and became simple portraits. Beginning as little more than a parlour amusement, the idea of these intimate snippets of self spread quickly across Europe and America, and across class and profession. They tapped into the growing enthusiasm for keepsakes and tokens of affection, offering a cheap, quick and accessible way to obtain a passable likeness. Itinerant silhouette artists sprang up in most fashionable cities and watering holes, snipping decent portraits from paper for a few pennies. The unfussy abstract nature of the image quickly became a way of expressing character in a direct manner, without the distraction of costume or setting, so much so that in the 1770s, the Swiss poet and philosopher, Johann Caspar Lavater, tried to establish the silhouette as a reliable scientific indicator of temperament as he worked on his studies of physiognomy. He called silhouettes 'the truest representation that can be given of man', relying on them in his hugely popular and influential *Essays on Physiognomy* (1789) which ran to many editions and many translations. By the early nineteenth century, the artist Augustin Edouart, working in London, was creating detailed expressive silhouettes of leading figures such as the actress Sarah Siddons and the writer Walter Scott, which captured a sense of movement and intense emotion in the simple dark portraits.

By the 1880s and 1890s, silhouettes had become a mainstay of art and theatre, with a growing alertness to the expressive potential of the simple form. The way in which silhouettes pare back faces and figures to essentials was being seen as a way of rendering internal truths rather than just outward appearance. Alongside his work with shadows, for example, Gaugin – along with other painters such as Paul Cézanne and Edgar Degas – explored the use of silhouettes to challenge the traditional way of modelling figures, working towards a more abstract representation that attempted to distil the essence of character. In his posters for theatre performances, Toulouse-Lautrec employed lively, distinctive silhouettes to capture the spirit of spectacle and of the stage personae. Taking this further, the experimental *théâtres d'ombre* or 'shadow theatre', popular in Paris in the 1880s and '90s, used flat silhouettes of figures cut from cardboard or metal to unsettling effect, conveying gesture, movement and emotion with striking directness. Borrowing from the ancient shadow theatres of South East Asia and the Middle East, these created ghostly figures of dark presence, which one contemporary critic suggested made audiences realise that 'all reality is only a reflection'.

Silhouettes proliferated despite (or perhaps because of) their austere, rather sombre qualities. People loved these little reminders of the dark. Although they might have begun as ephemeral amusement, as 'fast fashion', they quickly became subject to many of the superstitions around shadows and the colour black, and so weighted with more substantial meaning. They became tokens of loss, parting and death. Gaining particular resonance as *memento mori* and as private souvenirs of dead friends and lovers, they acquired a timelessness that belied their simple production and acted as a reminder of their extensive history. Long before Étienne de Silhouette had begun wiling away dull hours with scraps of black paper, silhouettes of deities cut from bronze sheets had been widely distributed throughout the empire of ancient Rome, perhaps used as inlays on musical instruments or furniture or as heirlooms, while Pliny tells the story of a girl who draws the silhouette of her departing lover on a wall, tracing the profile cast by the light of a lamp. Further back still, the earliest cave paintings often include silhouettes of black hands against the wall or pale hands against a smudged red or black background.

I first saw some of these silhouetted hands in the caves of Pech Merle, in the Célé valley in south-west France. There aren't very many of them. At Pech Merle most of the art is of animals, including striking dappled horses, but these slight, delicate hands stole the show for me. They are things of the dark. More than 2,500 feet underground, they would have been more or less invisible when they were first painted, faintly lit on occasion, mostly concealed by darkness. (Even today, for conservation reasons, they're lit only briefly.) Similar hands appear in caves from Indonesia to Argentina, from Borneo to Spain, in Africa and Australia, sometimes just a single silhouette, sometimes in a cluster, a crowd. The examples in France are among the oldest, but despite extensive study, we can't know what people meant by making a mark in this way. The hands are often female, but beyond this, we have no idea about them, whether they were painted for ritual or pleasure, in celebration or mourning, by eminent shamans or ordinary families. The silhouettes are immediately physical and personal, but there's no way of telling what they might have meant. A wave? A warning? A plea? Now they exist like the shadow in Hans Christian Andersen's story, independent of the people who made them, suggestive but unknowable. They reach out through time, a hand like my hand, an invitation into the dark; a shadow held in silhouette for over twenty thousand years.

The waning moon is a thing of fading shadows. The shadows of a full moon are dark and distinct, lively, splashed on the silver light like living things; as the cycle moves on, they grow slower and slighter, less defined.

It's New Year's Eve, mild, almost balmy. (The warmest New Year's Eve on record, apparently: we know this isn't a good sign.) I've spent the day with dad. He looked frail. His eyes were watery with what might have been tears, but he wore a party hat and sometimes laughed. He wouldn't eat, spending his time instead fingering the glitter of the seasonal decorations, chasing the light reflecting from the tinsel. It was a long drive home but now I'm back in the woods. The moon is a sliver of crescent, only a day or two away from the complete darkness of the new moon. It rises in the middle of the night, sliding between the branches of the trees like a cut toenail. It won't set until the afternoon – most likely you wouldn't notice it. Before it rises, the night is dark.

I can hear an animal approaching steadily, invisible, nothing more than a rustle of fallen leaves and the slight crackle of under-growth. A cat or fox, most likely. Something of that size. Slinky, quiet. It gives me a wide berth but doesn't hesitate. I hear it go on towards the stream.

The darkness hangs around me.

I hear something else now, the sizzle of distant fireworks, crackle-pop. But they do nothing to light this little valley beneath the trees.

I hear my breathing. The rub of my clothes. That is all there is.

A year slides, renews, under the slinking decline of the moon. I have walked a distance in the dark and come back here. I still don't know if I can find my way.

The night goes on. The shard of moon rides high. It feels as though the darkness moves over me with the passage of the night like a bolt of thick silk unscrolling. If I sit here until morning, it will finally have slid away. But of course it's me that's moving, moving into the night and beyond it. Night is the shadow of earth. This shadow remains still: it's always on the side away from the sun. The dark is always there for us, waiting. As the planet rotates, we spin into it.

Or perhaps that's the wrong way to look at it. Perhaps we begin in the dark, and rotate into the light and back again. Perhaps darkness is the natural state, the beginning and the end. We've fallen into the habit of thinking of daytime as the important marker. We organise time zones around the world to fit with the arrival and departure of daylight; we slide the clocks onto Daylight Saving Time to make the most of the long days. But I remember that Māori days begin at night; there are no days of the week, there are only nights of the moon. Looked at this way, the dark is afforded priority. It's not just something to be endured, slept through or ignored, but the fundamental structure around which the rest of life is arranged. This Māori night is so subtle and powerful, so full of potential, that it can't be contained by a single word – the strata and textures of darkness become explicit in a litany of names:

Te Po-nui (the great night); Te Po-roa (the long night); Te Po-uriuri (the deep night); Te Po-kerekere (the intense night); Te Po-tiwhatiwha (the dark night); Te Po-te-kitea (the night in which nothing is seen); Te Po-tangotango (the intensely dark night); Te Po-whawha (the night of feeling); Te Po-namunamu-ki-taiao (the night of seeking the passage to the world); Te Po-tahuri-atu (the night of restless turning); Te Po-tahuri-mai-ki-taiao (the night of turning towards the revealed world).

Great, deep, intense; seeking passage, nurturing feeling – the Māori terms conjure a sense of awe and reverence, the visceral reaction we have in the restless turning of the dark. As we've seen from the paintings in the caves at Pech Merle and elsewhere around the world, early humans lived at least part of their lives in the safety of darkness. Caves offered protection from the elements and from predators; they were a place of security and relative comfort, a site of domestic reassurance, of reverence and ceremony. We seem not to have lost this primal tolerance for the dark: it continues to provoke wonder and respect.

Have we brought with us, through all this time, an understanding that darkness is special? Are we, in the dark, utterly ourselves?

★⁺₊★ ☾★⁺₊★

There's not much comfort in dad's dementia. The disease means a long, bewildering, distressing decline, for those who live with it and those who care for them. I cannot share or relieve the dark that devastates my dad. I can't pretend to understand it. My investigation, on that basis, inevitably falls short, always inadequate.

But I see the dark now in a new way, more fully and more subtly. I see it in body and mind, in so much of our language and so many of our habits. And perhaps that's important, even if it's not enough. I think my dad would appreciate the effort.

The waning crescent eases through the sky towards dawn; the cycle of the moon has come full circle. It will all begin again, tomorrow, begin in darkness. 'No moon-talk at all now,' wrote the twentieth-century American poet, Carl Sandburg, 'Only dark listening to dark.'

I reach out my hand and I grope for dad in the dark. There's nothing to grab hold of. I'm here alone. In the pitch black. I am close to him and far away. The sticky dark holds around the two of us.

I leave the woods. I make my way home and turn on a light. Behind me, the dark closes.

I leave myself there. I leave my dad there. In fear and death, in solace and sanctity, in the unremarkable paradox that is this thing, this dark thing.

# NOTES

## 1. New Moon

'one may truly be said to see darkness…' John Locke, *An Essay Concerning Human Understanding*, 1690, book 2, ch. 8 sec. 3.6.

'visual tinnitus', Damon Rose, 'Do Blind People Really Experience Complete Darkness', *BBC News: Ouch*, 25 February 2015.

Anil Seth, *Being You: A New Science of Consciousness* (London: Faber and Faber, 2021), pp. 115 and 127. The quotations here only touch on the compelling and complex arguments proposed by Seth in this book and his numerous published papers exploring the evolution of our understanding of consciousness.

'the dead darkness…' in a footnote to Book 6, Georgius Agricola, *De Re Metallica*, translated by H. and L. Hoover, (London: The Mining Magazine, 1912), p. 218.

'I have to trap without a light…' evidence given by Sarah Gooder to the Ashley Commission on Mines (1842), *Parliamentary Papers*, 1842, vols. XV-XVII, Appendix I, pp. 252.

'what is the meaning…' John Ruskin, 'The Relation to Art of the Science of Light,' *The Eagle's Nest*. 2nd ed., (London: George Allen, 1891), pp. 114–37, p. 123.

Can we see in the dark? David Lewis, 'Veridical hallucination and prosthetic vision', *Philosophical Papers*, (Oxford: Oxford University Press, 1986), vol.2, pp. 273–86, p. 283.

'the distinctive and varying textures of things in the dark…' Alan Smith in the Nenthead mines, see 'Parameter Writings', Allanheads Contemporary Arts, Alan Smith parameter | acart-org-uk.

Account of Marietta Schwarz experiment, see for example, Will Hunt, *Underground: A Human History of the Worlds Beneath our Feet* (London: Simon & Schuster, 2019).

Sensory signals creating visual perceptions in the brain, see for example, Sekuler, R., Watamaniuk, S. N. J., & Blake, R., 'Perception of visual motion', in

H. Pashler & S. Yantis (Eds.), *Stevens' handbook of experimental psychology: Sensation and perception* (New York: John Wiley, 2002), pp. 121–76.

'My pictures are poetic objects capable of receiving what each person is ready to invest': Pierre Soulages qtd. on Christies.com, 'Beyond Black', 18 September 2019.

'a pure, abstract, non-objective, timeless …' Ad Reinhardt, see 'MoMA Learning' on MoMA | Ad Reinhardt. Abstract Painting. 1963.

'so emotive…' Jared Sexton, 'Basic Black', *Liquid Blackness: Journal of Aesthetics and Black Studies*, (Duke University Press), Vol. 5., No. 2., 2021, pp. 75–83.

'blackest forms of the disorder', Styon and Jamison qtd. in Femi Oyebode, *Sims' Symptoms in the Mind: Textbook of Descriptive Psychopathology*, (London: Elsevier, 2015) p. 263.

Adam Bartlett, 'Walking out of darkness', *Psychiatric Times,* vol. 36, no. 1 (January 2019).

Rats suffering neural damage in darkness, see, for example, Lisa Conti, 'How Light Deprivation Causes Depression', *Scientific American* (1 August 2008).

'When men in darknesse goe …' qtd. in Roger Ekirch, *At Day's Close: Night in Times Past* (London: Norton, 2006), p. 9.

For extensive and intriguing detail of medieval customs and rules regarding activity and dangers at night see Roger Ekirch, *At Day's Close: Night in Times Past* (London: Norton, 2006).

'It is in those drifting automatic states…' Stanley Hall, 'A Study of Fears', *American Journal of Psychiatry*, 1897, vol.8, no.2, pp. 147–249, p. 189.

'The darkness of the night reduces many…' A. W. Macfarlane, 'Notes on Distressing Awakenings', *The Lancet,* vol. 137, (11 April 1891), pp. 811–64.

Children's fear of the dark. See, Krisztina Kopscó and András Láng, 'Uncontrolled Thoughts in the Dark? Effects of Lighting Conditions and Fear of the Dark on Thinking Processes', *Imagination, Cognition and Personality*, Vol. 39, no. 1, September 2019, pp. 97–108, and Krystal M. Lewis, Kaushalendra Amatya, Mary F. Coffman, Thomas H. Ollendick, 'Treating nighttime fears in young children with bibliotherapy: Evaluating anxiety symptoms and monitoring behavior change', *Journal of Anxiety Disorders*, vol. 30, 2015, pp. 103–12.

'Accustomed to having a good footing in darkness …' Jean-Jacques Rousseau, *Emile: Or On Education*, translated with notes by Alan Bloom, (New York: Basic Books, 1979), p. 137.

'The oldest and strongest emotion of mankind is fear …' H. P. Lovecraft, 'Supernatural Horror in Literature', *The Recluse,* (1927), p. 1.

MRI scans of participants watching horror films, see M. Hudson, K. Seppala et al, 'Dissociable neural systems for unconditioned acute and sustained fear', *Neuroimage*, vol. 216, August 2020. https://doi.org/10.1016/j.neuroimage.2020.116522.

Responses to 'Bloody Mary' game. G.B. Caputo, 'Strange-Face-in-the-Mirror Illusion', *Perception*, vol. 39, 2010, pp. 1007–8.

Michael Myers in *Halloween*, Dave Portner, 'Don't Call John Carpenter A Horror Movie Director, Says John Carpenter', *Interview Magazine*, 2 February 2015.

'In Oslo the sun doesn't set until 10pm ...' 'Eskil Vogt', *Guardian*, 13 May 2022.

Dark and the startle reflex, C. Grillon, M. Pellowski, K. R. Merikangas, & M. Davis, 'Darkness Facilitates the Acoustic Startle Reflex in Humans', *Biological Psychiatry*, vol. 42, 1997, pp. 453–460.

'Surely it is more natural to think that darkness...' Edmund Burke, *A Philosophical Inquiry into the Origin of our Ideas of the Sublime and Beautiful*, Part IV, Section XIV, p. 141.

★⁺₊★ ☾★⁺₊★

## 2. Waxing

'You have forgotten that no resemblance exists ...' Bruno Latour, *What is the Style of Matters of Concern*, Spinoza Lectures presented to the University of Amsterdam, 2005, Lecture 2, p. 12.

'the red was first noticed...' G. M. Hopkins, 'The Remarkable Sunsets', *Nature* 29, 3 January 1884, pp. 222–3.

'If I could find a language in which to perpetuate those appearances...' Claude Lévi-Strauss, *Tristes Tropiques*, trans. John and Doreen Weightman (New York: Penguin, 1992), p. 62.

'the length of the twilight ...' Auguste Bravais, *Annuaire Métérologique de la France*, 1850, p. 34.

'the pinkish or purplish glow in the eastern sky ...' Margolle and Zurcher, *Les Meteores* (Paris: Hachette, 1869), p. 24.

'temperaments with a feeling for the sublime ...' Immanuel Kant, *Beobachtungen über das Gefühl des Shönen und des Erhabenen* (Rigam, 1771), p. 5.

Research on physical effects bearing out Kant's observations. See, for example, S. Mastandrea, S. Fagioli, V. Biasi, 'Art and Psychological Well-being: Linking the Brain to the Aesthetic Emotion', *Frontiers in Psychology*, 10, 2019.

Melancholy at dusk, Robert Burton, *The Anatomy of Melancholy*, ed. H. Jackson, 3 vols. (London: Dent, 1961), vol. 1, p. 254.

'although we cannot give any reason...' Kamo no Chōmei, qtd. in Earl Miner, Robert E. Morrell, Hiroko Odagiri, *The Princeton Companion to Classical Japanese Literature*, (Princeton: Princeton University Press, 1985), p. 12.

Influence of Circadian rhythms on mood etc, see, for example, Jeongah Kim, Sangwon Jang et al, 'Implications of Circadian rhythms in Dopamine and Mood Regulation', *Molecules and Cells*, Vol. 40, No. 7, July 2017, pp. 450–6.

Physical impact of circadian disruption: Ayesha Shafi and Karen Knudsen, 'Cancer and the Circadian Clock,' *Cancer Research*, vol. 79, No. 15, August 2019, pp. 3806–14; Ippei Shimizu, Yohko Yoshida, Tohru Minamino, 'A Role for Circadian Clock in Metabolic Disease', *Hypertension Research*, vol. 39, February 2016, pp. 483–91; 'Chronic Disruptions to Circadian Rhythms Promote Tumor Growth, Reduce Efficacy of Cancer Therapy – but how?', *Penn Medicine News*, 30 April 30, 2019; Brian Lananna and Erik Musiek, 'The Wrinkling of Time: Aging, inflammation, oxidative stress, and the circadian clock in neurodegeneration', *Neurobiology of Disease*, vol. 139, June 2020, open access online.

Light level effects on hormone production: see, for example, T. A. Bedrosian and R. J. Nelson, 'Timing of Light Exposure Effects Mood and Brain Circuits', *Translational Psychiatry*, vol. 7, No. 1, January 2017.

External light pollution affecting sleep: Korean research published in *Journal of Clinical Sleep Medicine*, December 2013, and November 2018.

Effect of light pollution on songbirds: Arnaud da Silva, Mihai Valcu, Bart Kempenaers, 'Light Pollution alters the Phenology of dawn and dusk singing in common European songbirds,' *Philosophical Transactions of the Royal Society*, May 2015, https://doi.org/10.1098/rstb.2014.0126.

Sand hoppers, and light pollution effect on coastal areas: Daniela Torres, Svenja Tidau, Stuart Jenkins, Thomas Davies, 'Artificial Skyglow Disrupts Celestial Migration at Night', *Current Biology*, Vol. 30, No. 12, June 2020, pp. 696–7.

Garcia Lorca's description of the *duende*: 'Theory and Play of the Duende', translated by A. S. Kline, *Poetry in Translation* (2007).

'the discovery of dusk…' Mark Cocker, *Crow Country* (London: Vintage, 2008), p. 152–5.

Ghosts 'embittered the lives of a great number of persons…' Francis Grose, *A Provincial Glossary*, (London: Edward Jeffrey, 1811), p. 242

'In Scotland last year while walking through an ancient forest…' *Country Life*, 27 February 1942, and reprinted 27 October 2021.

'simple for spectres to appear…' Jun'ichirō Tanizaki, *In Praise of Shadows* (London: Vintage, 2001), p. 56.

'Have you ever felt a sort of fear…' ibid., p. 35.

## 3. Full Moon

For discussion of the popularity of almanacs, see Roger Ekirch, *At Day's Close, Night in Times Past* (London: Norton, 2006), p. 129.

'In the country, the darkness of night is friendly…' W. Somerset Maugham, *A Writers Notebook* (London: Heinemann, 1951), p. 37.

'How much more mysterious…' Juhani Pallasmaa, *The Eyes of the Skin: Architecture and the Senses,* (London: Wiley Academy, 2005), p. 50.

'The history of architecture…' Le Corbusier, *The Radiant City: Elements of a Doctrine of Urbanism to Be Used as the Basis of Our Machine-Age Civilization* (London: Faber, 1967), p. 71.

'Greek architecture taught me that the column is where the light is not…' qtd in 'Greek Architecture and Light / Louis Kahn', *ArchEyes Magazine,* 20 February 2020.

Research project by the Helen Hamlyn Centre for Design, published in *Light Volumes Dark Matters* (Royal College of Art, 2010), p. 28.

'I came to the mouth of a great cavern…' qtd. in Walter Isaacson, *Leonardo da Vinci* (London: Simon and Schuster, 2017), p. 20.

Candlelit procession at Walsingham, reported in the *Norwich Mercury,* reproduced in *Our Lady's Mirror,* Autumn 1928.

'Whiro was driven down to the underworld…' 'Māori personifications. Anthropogeny, solar myths and phallic symbolism: as exemplified in the demiurgic concepts of Tane and Tiki', *The Journal of the Polynesian Society,* vol. 32, no. 2, (1923), p. 107. Whiro is now more commonly and accurately named Whiro-te-tipua

'the unformed and void…' Genesis 1: 2.

'people could not see one another …' Exodus 10:21–3

'the light that shines in the darkness,' John 1:5

'the light has come into the world…' John 3:19–20

'the eye is the lamp of the body…' Matthew 6:22

'Our congregations have been scattered…' Huguenot prayer, qtd. in Riolf Reichardt, 'Light against Darkness: the Visual Representations of a Central Enlightenment Concept', *Representations,* Vol. 61, Winter 1998, pp. 95–148, p. 104.

Accounts of Sufi ritual on the Ivory Coast, André Chappette, 'When silence is "yeelen" (light): Exploring the corporeality of the mind in a nocturnal solo *zikr* practice (Odienné, Ivory Coast)', *Critical Research on Religion,* vol. 9, No.2, pp. 175–90.

'I must have a dark side …' Carl Jung, 'Practice of psychotherapy: Essays on the psychology of the transference and other subjects,' *Collected works of C. G. Jung,* vol. 16. (Princeton, N.J.: Princeton University Press, 1966), p. 59.

'pure light and pure darkness are two voids …' George Hegel, *The Science of Logic,* Vol. 1, p. 93.

Voltaire, 'pass through the darkness of ignorance and lies', *Romans et Contes,* ed. H. Benac, (Garnier frères: Paris, 1960), p. 487.

'Pseudoscience and superstition will seem year by year more tempting …' Carl Sagan and Ann Druyan, *The Demon-Haunted World: Science as a Candle in the Dark* (New York: Headline, Random House, 1996), p. 29.

'Despite the obscurity of the twilight …' Baron de Holbach, *Essai sur les Préjugés,* qtd in Riolf Reichardt, 'Light against Darkness: the Visual Representations of a Central Enlightenment Concept', *Representations*, Vol. 61, Winter 1998, pp. 95–148, p. 110.

'the rubbish of the dark ages', Edward Gibbon, *The History of the Decline and Fall of the Roman Empire*, vol. 6, ch. 37, pgh. 619.

'Let the dark benighted pagan'. The original first line of the second verse of this hymn, known as 'O'er the Gloomy Hills of Darkness' took in a wide sweep of world cultures. Congregations sang 'Let the Indian, let the Negro', until this was changed by enslavers in the USA to 'Let the dark benighted pagan', with its more overtly religious overtones.

'the light of the gospel ...' William Ellis, *Madagascar Revisited, Describing the Events of a New Reign and the Revolution Which Followed*, (London: John Murray, 1867), p. 459.

'I feel disposed to think there must be a physical and, consequently, a psychological inferiority in the dark races generally', Robert Knox, *The Races of Men: A Philosophical Inquiry into the Influence of race over the Destiny of Nations*, (London: Renshaw, 1850), p. 224.

Elizabeth Arden advertisement. *Vogue*, 1 July 1920, p. 112.

'greater whiteness and brilliancy...' Arnold Cooley, *The Toilet and Cosmetic Arts in Ancient and Modern Times* (London: J&A Churchill, 1872), p. 428.

For the hierarchy of race by Carl Gustav Carus see, for example, Milan Hrabovský, 'The Concept of "Blackness" in Theories of Race', *African and Asian Studies*, Vol. 22, No. 1. 2013, pp. 65–88, p. 82.

'a real or fabricated Africanist presence ...' Toni Morrison, *Playing in the Dark: Whiteness and the Literary Imagination*, (Harvard University Press: Cambridge, Mass. and London, 1992), p. 6.

'the projection of the not-me', ibid. p. 38.

Refugees as a 'dark force', *Daily Mail*, 16 February 2016, 4 July 2017.

Studies in press photographs and race experiments by New York University: Adam Alter, Chadly Stern, Yael Granot, Emily Balcetis, 'The "Bad is Black" Effect: Why People Believe Evil-doers have darker skin than do-gooders,' *Personality and Social Psychology Bulletin*, Vol. 42, No. 12, November 2017, 1653–65.

Missionary diary discussing Māori women, Kathryn Rountree, 'Re-Making the Maori Female Body: Marianne William's Mission in the Bay of Islands', *The Journal of Pacific History*, Vol. 35, No. 1, June 2000, pp. 49–66, qtd p. 60.

'fiery' and 'hot constitution'd' African women, in D.G. White, *Ar'n't I a woman? Female slaves in the plantation South* (Rev. ed.) (New York: Norton, 1999), p. 29.

The Hottentot Venus, 'Law Report', *The Times*, 26 November 1810, qtd. in *The Times Digital Archive*, 7 August 2012.

Description of lovers in Rome, *Town and Country*, No. 24, 1792, p. 261.

'what is held up as the story of the feminine soul', Margaret Oliphant, qtd in Susan David Bernstein, 'Dirty Reading: Sensation Fiction, Women and Primitivism', *Criticism*, Vol. 36, No. 2, Spring 1994, pp. 213–41, p. 213.

'The sexual life of adult women is a "dark continent" for psychology', Sigmund Freud, 'The Question of Lay Analysis', *The Standard Edition of the Complete Psychological Works of Sigmund Freud,* vol. 20, pp. 177–258, p. 212.

'Sex gradually became an object of great suspicion …', Michel Foucault, *The History of Sexuality,* vol. 1, p. 69.

'this darkness, this mystery, this intrigue…' Johnny Depp, in 'Tim Burton Interview – Dark Shadows', www.moviesonline.ca. 23 November 2017.

'In all the best erotic writing …' Niamh Campbell in the *Guardian,* 14 February 2022.

'darkness wrapped around her tenfold…' Jun'ichirō Tanizaki, *In Praise of Shadows* (London: Vintage, 2001), p. 53.

'the darkness might be of the valley of the shadow of death…' Edward Bradbury, *Midland Railway Sketches,* (Sheffield: Midland Railway Society, 1999), p. 48.

'flight through the realms of chaos…' *Drake's Road Book of the London and Birmingham Railway,* (London: Hayward and Moore, 1839), p. 14.

'we are entangled in, and appropriated by…' Carl Mika, 'Subjecting ourselves to madness: A Māori approach to unseen instruction,' *Educational Philosophy and Theory* vol. 53, 2021, pp. 719–27, p. 719.

## 4. Waning

'Now that the darke has come…' is a prayer advocated for nightly recitation in W. F. (William Fiston), *The Schoole of Good Manners or, A New Schoole of Vertue* (1609). Printed by I. Danter, for William Ihones: and are to be sold at the signe of the Gun neare Holburne Conduit., 1595.

Antje Ravic Strubel, 'We Tell Ourselves Stories', in *Death in Literature,* ed. Outi Hakola and Sari Kivistö (Cambridge Scholars Publishing: Newcastle Upon Tyne), 2014, pp. 3–14.

'I felt the aloneness, the emptiness of space…' qtd in 'Distressing Near-Death Experiences: The Basics', Nancy BChaush and Bruce Greyson, *Missouri Medicine,* Vol 111. No. 6, Nov–Dec 2014, pp. 486–91.

'I went into a dark place…' qtd. in *'Tales of the Dying Brain'*, Christof Kock, *Scientific American,* Vol. 322, No. 6, June 2020, pp. 70–5.

'floating towards a dark…' qtd. in 'Near-Death Experiences: Evidence for their reality', Jeffrey Long, *Missouri Medicine,* Vol 111. No, 5, Sep-Oct 2014, pp. 372–80.

'there was only darkness…' qtd. in 'What do near-death experiences mean and why do they fascinate us?', Alex Moshakis, *Guardian,* 7 March 2021.

Hibernation more akin to sleep deprivation than sleep. 'Could Humans Hibernate?', Vladyslav Vyazovskiy, *The Conversation,* 15 March 2016.

The account of Russian 'hibernation', quoted in *British Medical Journal,* 6 May, 2000.

Research into winter effects on scientists at polar research stations, 'Psychological Hibernation in Antarctica', Sandal, van de Vijver and Smith, *Frontiers in Psychology*, November 2018, https://www.frontiersin.org/articles/10.3389/fpsyg. 2018.02235/full

Details of REM and NREM sleep, US Institute of Sleep Medicine and Research, *Sleep Orders and Sleep Deprivation: An Unmet Public Health Problem* (Washington: National Academies Press) 2006.

Positive effects of REM sleep. I. Lerner, S. Lupkin, N. Sinha, A. Tsai, and M. Gluck, 'Baseline Levels of Rapid Eye Movement Sleep May Protect Against Excessive Activity in Fear-Related Neural Circuitry', *Journal of Neuroscience,* Vol. 37, No.46, November 2017, pp. 11233–44.

'acoustic intimacy', Juhani Pallasmaa, *The Eyes of the Skin: Architecture and the Senses,* (London: Wiley Academy, 2005), p. 50–3.

'I am developing the art of gazing with my hands…' John Hall, *Notes on Blindness: A Journey Through the Dark,* (London: Wellcome Collection, 2017), p. 155, p.172–3.

'the most necessary, noblest, clearest and most subtle member above all others' qtd in Stuart Clarke, *Vanities of the Eye: Vision in Early Modern European Culture* (Oxford: Oxford University Press, 2003), p. 10. This book offers a useful discussion about the primacy of sight, and the development of theories of vision.

'putrid… stinking like death, like sin.' Alain Corbin, *The Foul and the Fragrant* (Cambridge, Massachusetts: Harvard University Press, 1986), p. 143.

Anne Carson, 'The Gender of Sound', *Glass, Irony and God* (New York: New Direction Books, 1992), pp. 119–42, p. 119.

Mice and human ability to see shadow, Aalto University, 'The Limits of Vision, Seeing Shadows in the Dark', *Science Daily,* 23 May 2022.

Buzz Aldrin qtd. in Tony Philips, 'Apollo Chronicles, Dark Shadow', NASA, 30 June 2001.

'Were it not for shadows there would be no beauty.' Jun'ichirō Tanizaki, *In Praise of Shadows* (London: Vintage, 2001), p. 46.

J.M. Barrie's 'Fairy Notes', qtd. in Swank, Kris, 'The Shadow-Self and Coming of Age in George MacDonald's *Phantastes*, Ursula Le Guin's *A Wizard of Earthsea*, and J. M. Barrie's *Peter Pan*', (2015), p. 9, available on Academia.edu

Paul Gaugin, 'that if, instead of a figure you put the shadow only of a person, that is an original point of departure', qtd in Nancy Forgione, 'Shadow Only: Shadow and Silhouette in late nineteenth-century Paris', *The Art Bulletin*, Vol. 81, No. 3, pp. 490–512, p. 490.

'morally reprehensible tendencies' and discussion of the shadow as 'that hidden, repressed, for the most part inferior and guilt-laden personality whose

ultimate ramifications reach back into the realm of our animal ancestors…',
Carl Jung, *Collected Works*, 9, part 2, paragraph 422.

Johann Caspar Lavater, 'the truest representation that can be given of man',
*Essays on Physiognomy; for the promotion of the knowledge and the love of mankind*,
(Boston: William Spotswood, 1794) p. 291.

'All reality is only a reflection', Nancy Forgione, ibid, p. 502.

Ancient cave paintings of silhouetted hands largely female, Virginia Hughes,
'Were the First Artists Mostly Female', *National Geographic*, 10 October 2013.

*Māori names for the dark*, qtd in Amadonna Jakeman,. 'Hei Poai Pakeha koutou
i muri nei. You Shall Be Pakeha Boys. The Impact of Te Tangi O Kawiti on
Ngati Hine Resistance to the Crown in the Treaty Claims, Mandate and
Settlement Process', PhD diss., Auckland University of Technology, 2019, p. 33.

## ACKNOWLEDGEMENTS

Thanks to everyone who was willing to talk about what the dark meant to them, and to own up to being afraid of it. Thanks also to everyone who has helped me by sharing bits and pieces of specialist knowledge about the dark. Several colleagues have been generous in recommending material: particular thanks are due to Alice Vernon for recommending some useful sources and to Jowhor Ile for invaluable help in guiding and challenging my discussion of race. Thanks to Paul Dale for a pleasant canalside stroll mulling over meanings of the dark in Christian tradition. Working with Icon Books has been a pleasure. I'm particularly grateful to Ellen Conlon for her enthusiastic response to my proposal and Sophie Lazar for such supportive and incisive editing. Most of all, thanks to mum for encouraging me to write this book so that we might all understand dementia a little better and approach it with kindness.

# INDEX